## Praise for *Brilliant Workshops*

Cyrus has a wealth of experience in designing and delivering learning and development solutions. The key to his success is how he works collaboratively with his clients to truly understand their needs and always having the courage to be himself. This book will help you develop your own training style and enable you to deliver workshops that have a unique personal touch and add value to business.

*Elizabeth Crosse, Head of Learning and Development,*
*Legal Services Commission*

Cyrus Cooper has had an ongoing relationship with Thomson Reuters for a number of years focusing on the training and development of our graduates. His experience of creating and delivering workshops that add value to attendees and drive results is vast, and the results speak for themselves; our graduates come out more secure in themselves, more prepared for working life and with an array of soft skills they can use every day. This book will be an excellent bible for any manager who wishes to develop similar workshop results.

*Elizabeth Pedler, Head of Graduate Programmes, Markets Division,*
*Thomson Reuters*

Cyrus Cooper connects quickly and effectively with his audiences. Along with excellent content, this provides a powerful platform for brilliant training to occur. In *Brilliant Workshops*, Cyrus shares with the reader many and much of what has made him and can equip you to deliver brilliant workshops of your own.

*Ken Buist, Founder of The Trusted Adviser, Author and Speaker*

D0307538

Cyrus Cooper has worked with our organisation for the past six years, creating and delivering an array of workshops and facilitating team development sessions. His calm approach has drawn positivity from the toughest of audiences and he seems to arrive at the heart of the issue with ease, leading the most difficult characters towards solutions. His insights are astute and his advice is worth following. Above all? He's a genuinely nice bloke.

*Katy Grêlé, HR Director, Legoland Windsor*

**brilliant**

# workshops

# PEARSON

At Pearson, we believe in learning – all kinds of learning for all kinds of people. Whether it's at home, in the classroom or in the workplace, learning is the key to improving our life chances.

That's why we're working with leading authors to bring you the latest thinking and the best practices, so you can get better at the things that are important to you. You can learn on the page or on the move, and with content that's always crafted to help you understand quickly and apply what you've learned.

If you want to upgrade your personal skills or accelerate your career, become a more effective leader or more powerful communicator, discover new opportunities or simply find more inspiration, we can help you make progress in your work and life.

Pearson is the world's leading learning company. Our portfolio includes the Financial Times, Penguin, Dorling Kindersley, and our educational business, Pearson International.

Every day our work helps learning flourish, and wherever learning flourishes, so do people.

To learn more please visit us at: www.pearson.com/uk

# brilliant

# workshops

How to deliver effective workshops to any audience

Cyrus Cooper

**PEARSON**

Harlow, England • London • New York • Boston • San Francisco • Toronto • Sydney • Auckland • Singapore • Hong Kong
Tokyo • Seoul • Taipei • New Delhi • Cape Town • São Paulo • Mexico City • Madrid • Amsterdam • Munich • Paris • Milan

**PEARSON EDUCATION LIMITED**

Edinburgh Gate
Harlow CM20 2JE
Tel: +44 (0)1279 623623
Fax: +44 (0)1279 431059
Website: www.pearson.com/uk

First published in Great Britain in 2012

ISBN: 978-0-273-75975-1

*British Library Cataloguing-in-Publication Data*
A catalogue record for this book is available from the British Library

*Library of Congress Cataloging-in-Publication Data*
Cooper, Cyrus.
  Brilliant workshops : how to deliver effective workshops to any
  audience / Cyrus Cooper.
    p. cm.
  Includes bibliographical references and index.
  ISBN 978-0-273-75975-1 (pbk.)
  1. Workshops (Adult education) I. Title.
  LC6562.C66 2012
  658.3'124--dc23

                                        2011036366

The publisher is grateful for permission to reproduce Figure 6.1 adapted
from *Project Management: A Systems Approach to Planning, Scheduling, and
Controlling*, 10th Edition, by Harold Kerzner. Copyright © 2009 John Wiley &
Sons, Inc. Reproduced with permission of John Wiley & Sons, Inc.

10 9 8 7 6 5 4 3 2 1
15 14 13 12 11

Typeset in 10/14pt Plantin Std by 3
Printed and bound in Great Britain by Henry Ling Ltd, at the Dorset Press,
Dorchester, Dorset

*This book is dedicated to my wonderful family – Jo, Jess and Olly – for giving me the time, support and encouragement to do the things I do, and who are everything to me*

# Contents

# About the author

**Cyrus Cooper** has had a different career route from many in the world of facilitating. His dream as a young boy was to be in a rock band. Music was his passion and he was in many bands in his late teens to early twenties – a world away from the learning and development field you may think? Not so. Being in a band requires you to be onstage and in front of an audience. You have nerves before you go on. You want to make that incredible first impact on the audience. You constantly test out how well you are being received. You are conscious of the potential performance dip half-way through. You go for the high impact ending. Is the audience satisfied? Will they want you back?

A great performance is required each time you take the stage. Whether you have performed the same songs every night – the audience want that same enthusiasm and passion. For trainers and facilitators, this sounds familiar. Therefore, this book has been written with deliberate parallels to his band experience.

After quickly working his way up through local government and the Civil Service, he got the studying bug and holds a master's degree in Management Practice, as well as diplomas in leadership, training and corporate and executive coaching. Cyrus is also accredited to use a variety of tools that assist development in a multitude of areas.

# Acknowledgements

There are so many fantastic people that I have worked with over the years. I have been influenced by their contagious positive energy. You know who you are. However, I do want to make particular mention to Ana Antunes da Silva, who has supported me throughout this whole book-writing adventure and been honest with me when I needed it most.

My music has been my source of inspiration. Without Adam and the Ants and The Clash, I would never have got onstage myself, which may have meant I would never have found a job I absolutely love doing, being in front of people.

We can learn a lot from how actors prepare before appearing onstage or on the television. I would like to thank my good friends at Silvercube (**www.silvercube.co.uk**), particularly Steve Leslie, who excel in 'actor inspired communication'. They have contributed to sections of this book.

Also, to my trusted colleagues, Michael, the two Jos and Linda. And Graham when you pop in. You make it a great place to work. Keep the latte's coming…

Finally, I would like to thank Eloise, my editor. I have learnt so much about writing styles and the impact a single word can make. Your feedback has been both insightful and prompt – how do you read so fast? This has really made it a powerful learning experience for me. Thank you so much.

# Foreword

I was delighted when I asked Cyrus if he had a foreword written for his book and he said 'not yet' – I seized my chance and offered and he kindly accepted. This is not a preamble to a 'lifetime achievement award' at the BAFTAs, but I will start with when I first collaborated with him.

It was in the heady days of a new leadership framework and performance management system being introduced into a newly enlarged (through merger) government department. We both volunteered and were allocated the responsibility of delivering a range of modules, but these were training courses, so not purely facilitation. We delivered many courses together – workshops in which we were getting to the heart of the issues, the needs and the challenges we would face in running those workshops. It was then I realised I had a 'partner in crime' in facilitation. Cyrus was adept in that space in so many ways: settling people down and building a light but professional air to the proceedings; spurring people on to contribute; capturing thoughts and turning them into provocative discussion points; driving to conclusions; mixing up the formats to stimulate and energise; using humour to create a liveliness to the room and keeping to time. In fact, I credit this man with keeping me to time in a classroom. I was *always* overrunning until I started working with Cyrus. And time is so important to us all – so overrun at your peril these days!

When our career paths took us to different places, I was delighted when he was able to work with me as a consultant. Again his design and training course delivery prowess was in full demand and maximised, but it was his facilitation that my internal clients still remember the most. The ad-libbed briefs he worked up, the challenging management team events he drove forward to productive outcomes – it's this reason why on LinkedIn I referred to him as the 'most internal external' I've ever worked with. He got the culture so much that facilitation was not a job to him, it was something he emotionally, intellectually and professionally connected to.

So in summarising Cyrus, he's the man who makes the facilitation game his own. Why? He's interested, he uses all he knows and he creates. He creates dialogue, energy and solutions – which to me are the Holy Trinity of facilitation. The skills required to run your own brilliant workshops are all here.

It's great to see Cyrus's skill and passion for facilitation in the words used and thoughts conjured up in this book – brilliant facilitation beckons and in Cyrus we trust.

*Perry Timms*
*Head of Talent and Organisational Development, Big Lottery Fund*

# Introduction

At some point in your life, you will attend a workshop. You will also be asked to speak in front of others. Some people revel in the role and others are scared to death! Whatever you feel about speaking in public, it is an essential skill that is admired by others. We also all get involved in training at some point, whether as a trainer or facilitator or as a participant. Our experiences will greatly differ. This book is for people who are new to the subject, want a quick fix or have little or some experience in delivering workshops. It will look at the best fit for the role of 'facilitator' or 'trainer' as well as delivering to different types of audience. You will benefit from reading this book if you are:

- a learning and development (L&D) professional;
- a general manager – you may be recently promoted and now have a team to manage – and the organisation now requires you to 'train' your team;
- a freelance trainer;
- a consultant;
- a supervisor – who needs new ways of getting your team to improve its performance.

I personally remember being asked to 'run a session' at a staff training day at Hendon Magistrates' Court. I was relatively new and obviously wanted to impress my boss as I was also eyeing promotion. The way it was 'sold' to me was that it was just a short session and that I should find it 'no problem'. As the date drew nearer, I was getting anxious, nervous and feeling out of my depth. I knew nothing of planning my part of the day and where it fitted in, thinking about who the audience was, how I would structure my session and how I would talk in front of 60 people. This fear was very real. My big learning from that day was that when people say, 'You'll be fine...', I would question it and really understand what their expectations are.

If you can associate yourself with these feelings and you could do with a little support, then this book is for you.

You may be starting off in learning and development (L&D), or already have some experience in designing and delivering workshops. You may be a manager who manages staff and you are asked to run training sessions for them – not necessarily because you are a good trainer/facilitator, but because you are their manager.

Whether you are employed internally or externally, this book is all about giving you the tools to run brilliant workshops. It is very practical, to help you see and understand what you need to do to be effective. The focus is not on theory, but rather on practical language and explanations that will enable you to run that workshop next week. It will give you bulleted checklists that will enable you to make quick decisions.

With 17 years' experience running workshops, both as an internal trainer and latterly as an external consultant, I feel that I am in a position to share what has brought me success, by looking at what I could have improved and what I did well. This book is weighted in favour of experience and will have you wanting to 'crave the feedback' – to enable you to focus on what people want and for you to continuously improve. You will be nodding your head at the examples given as you will have been there – or know someone who has. It follows a clear pathway from the initial request for 'training' to effective evaluation that makes you feel that you could be happy making a living out of this.

I want to show that you do not always have to be a subject expert, or the most senior person present, before you can be placed in front of a group or team. If you are facilitating, you need to have confidence, but your skills should be all about people, understanding group dynamics, how people learn and what you can do to enable them to walk away achieving what they wanted to achieve. You can be the new CEO – Chief Enabling Officer!

## The themes

### Part 1 Backstage (pre-workshop)

You have practised, planned and built up your confidence. You know you have done everything in your control to get you to this point. The five Ps (see Chapter 1) is a great technique for me, but simply knowing all the acronyms and letting your audience know you do will not enable you to deliver a brilliant workshop. This section will focus on what to think about when engaging with the person requesting the learning intervention (the 'client') and how to build on it in the design. We need to be smart about finding out what they want and not assume that they know what they want.

### Part 2 Onstage – stand and deliver (during the workshop)

No, we don't have to be in Adam and the Ants to do this part (for those of you who remember their hit song), but in effect we are onstage, performing to an audience, getting a 'feel' for how it's going and adapting as we go along to make sure it is memorable and has the desired effect. We can also be self-conscious, and if we play a wrong note we hope the audience doesn't notice – they mostly won't, unless we stop and tell them! We need to manage the audience as a whole, as well as individuals. Tactics and strategies to help you manage group dynamics will help you to overcome any nerves and get you in the right frame of mind to deliver.

## Part 3 Encore (post-workshop)

Will you get asked back? What memories will you leave behind? Without doubt, you can make a difference. Workshops can 'change' people for the better. I have seen how powerful they can be. I have had people follow their dream job that they thought was unattainable, have that difficult conversation they had been putting off for years, managing upwards with confidence, standing up and talking in public. The list goes on. We will look at what type of legacy you want to create. You can achieve great things for others. **YOU** are the enabler.

# PART 1

# Backstage (pre-workshop)

## How do I get clarity?

Chapter 1 will help to identify exactly what your client wants. That may not be the same as what they need though. This chapter arms you with the questions and techniques to differentiate between want and need.

## Holding up the mirror – your strengths and development areas

Chapter 2 enables you to take a good look at yourself. Yes, this bit can be painful, but it is essential if we are to deliver for our client. What am I good at? How do others see me? Where can I develop? Do I need to tell the audience what they need to do or do I need to facilitate a group of people to get them to reach their goal? Getting ready will pay dividends for you.

## How do I begin?

Chapter 3 gives you some tools to help you in making those first workshop plans. You can use what is here or design your own. This stage is crucial. It may not be as exciting or as nerve-wracking as standing up in front of an audience, but knowing you are fully prepared can give you the extra confidence to do it.

## I need to know who is coming

It is all about the audience. Chapter 4 gets into investigating all we can about them. Why are they here? Do they want to be here? What if they want to be somewhere else? If we can look at questions like this and answer them in advance, it can give us an understanding that feeds into our chosen approach and style.

## I need to have a range of options

Chapter 5 looks at some tools you may already use, or want to use, to support some of your workshops. You don't have to use any of them. It depends on whether or not you want to delve deeper into facilitation and build up your knowledge to offer your clients more choice.

## Do I print out workbooks or handouts? Are they needed?

Chapter 6 considers what resources the participants need to enhance their learning. We look at the pros and cons of all different types of resources and whether to print, email or just rely on participants' memory to remember everything we have said.

## How can I make a positive impact?

Chapter 7 rounds off our backstage plans. Think of your appearance and whether to get rid of all the chairs and just have seats. How will you set up the room? What is going to work for this specific group of people? Are you starting with a high tempo introduction or a softly spoken one?

There's lots to think about, so let's begin the journey...

**CHAPTER 1**

# What is wanted? What is needed?

You knew this day was coming. You need to be ready and use the five Ps:

- Proper
- Planning
- Prevents
- Poor
- Performance.

This chapter will focus on truly understanding the client. A request has come your way and you now have to become a mind-reader to decipher the training requirement. You will ask brilliant questions to get the answers you want. You will engage effectively and realise that you will need to get the workshop right to reach the right result. It will focus on want vs need and how you can begin to take control of the situation. Question whether training is the right solution for them.

 You can tell whether a man is clever by his answers. You can tell whether a man is wise by his questions.

Naguib Mahfouz

## Defining the need

### The first contact

You may receive many different requests for 'training'. It can range from a quick phone call to a long consultation meeting. I remember coming into my office to find a Post-it note requesting management training for a team. There were four small lines of text, including the date – three days' time. Talk about a challenge. A great starting point is to understand the difference between a trainer and a facilitator. Clients will often use both terms without differentiation. Although there are many similarities, it is the differences that you should be aware of.

More importantly, the requester (or client as I will refer to them throughout) needs to know the difference between training and facilitation. This will unpack the required role for all parties.

 **definitions**

**Trainer**

A trainer is 'someone who trains people'.

**Facilitator**

A facilitator is there to 'make (an action or process) easy or easier'.

(*Oxford English Dictionary*)

### Facilitation

 The facilitator's role is much more about ***opening things up for discussion*** in a stimulating way, getting ideas into the open and helping the group to listen to each other, further its knowledge and thus make informed decisions.

Cameron, 2001

A trainer starts from their own knowledge – whereas the facilitator starts from the group's knowledge. Ideally, they should be process leaders only – they have no decision-making authority and nor should they contribute to the substance of the discussion. To be effective as a facilitator, the role should, ideally, be a neutral one. There may be no right answer, but the skill is in guiding the group to their preferred outcome.

Often, managers 'facilitate' workshops within their team but cannot remain truly neutral as the outcome will affect them in some way. This can find them leading the conversation, consciously or sub-consciously, to a preferred outcome. A facilitator's role can be compared to that of a football referee. It is often said that the best referees are the ones who have rarely been heard and have allowed the game to flow. A facilitator can be successful just by being an enabler. A facilitator does not have to be heard the most or be seen the most to be successful.

The majority of facilitators are subject knowledge specialists who have the skills to transfer that knowledge through a range of activities that are best suited to the learners. They will tend to have an interactive style, mingling through the participants, having some one-on-one discussions along the way and seeing themselves as part of that group.

## Training

Training is more about directing what needs to be done. There is little room for manoeuvre or discussion. Trainers tell participants how it should be, what the ideal is and why they need to do it. They want participants to know the *right* answer. They may not get buy-in; they may not care, because following a particular process may have to happen for the team or organisation to succeed. For example, if an organisation wants to roll out a new appraisal system to hundreds or thousands of people, a trainer who understands the process, time-cycles and forms to be used will be suitable, as this is what every member of staff must do. There will be no alternative option.

Working in the Civil Service, part of my role was to do just that and it can become repetitive. However, it can also build confidence in the subject area, and by allowing no room for alternatives you can clearly and strongly set the tone and parameters for the expectations. A trainer will tend to deliver from the front of the room and will do most of the talking. There are times to train and times to facilitate. Knowing when to do this will ensure that your style matches the situation.

 I cannot stand being taught – but I enjoy learning.

Sir Winston Churchill

## The initial request

When you receive a request for training, you must never assume the client really knows what they truly want or need. This is where your questioning skills will be vital. Do not be afraid to ask many questions. For the workshop to be the best it can be for the client you need to truly understand as much as you can about what is required. Ask the person who has made the initial request if they are affected by the outcome. They may just be a messenger and may not be able to answer your questions accurately.

Understanding the need is one of the most important conversations you will have with the client. Invariably, experience tells me that if the training does not 'hit the spot' it is because the brief was not fully explained or understood. This is why we need to get as much information as possible and it is up to us to get clarity. If it is your chosen/current career, for credibility you need to be seen as the specialist, as there may be a lot riding on the outcome. If the client cannot articulate what success looks like, for example, then you should work with them to unpack this, not go away and 'have a go'.

If, for example, you are facilitating a teambuilding workshop, why does the team require teambuilding? Are there sensitivities? Does X get on with Y? Is teambuilding just a title to cover up other issues? Is it a performance issue that sits better by calling it teambuilding? This last point is a common 'misunderstanding'.

Think twice before agreeing to run workshops if you are not sure what a successful outcome looks like. Your reputation is on the line so you need to get clarity. The way I remember this is:

**C**larity **Y**ields **R**esults for **US**.

This statement of intent remains in my thoughts as it spells out my first name. Look for triggers that work for you.

A common acronym is SMART. This is used for goals and objectives.

- **Specific**: Objectives/goals must be clearly defined. For example, to deliver a dynamic presentation, it must be clear what makes a dynamic presentation.

- **Measurable**: Objectives/goals must be clearly measurable in terms of quality, quantity and time. This approach will help both you and the client assess progress at all times.

- **Achievable**: Objectives/goals must be realistic. Objectives/goals that are consistently set too high can become frustrating. On the other hand, when objectives/goals are set too low, the person achieving them may not get any satisfaction.

- **Relevant**: Working with the client, ensure that any objectives/goals take into account a link to a bigger picture. This will make sure that work is focused on achieving the necessary results. Alternatively, the client may just want a fun event as a reward for the team. Always check the context in which the workshop will be delivered.

- **Timebound**: Objectives/goals should always include timescales. If they didn't, it would make it very difficult to plan ahead. Aim to include as many milestones as you can. Remember – no date, no good!

 **timesaver**

Always ask the client what a successful workshop would look like to them. You should be able to get them to define this. A good way of doing this is getting them to imagine it is the end of the workshop and all the participants are telling them that it was a really great day, everyone was committed, and maybe everyone had fun while learning. It was also so useful for them that they could really make a positive difference to the team. With all of the above, make sure they are specific and measurable.

A quick four-stage process (see Figure 1.1) will help you to focus on delivering content that is relevant for the audience.

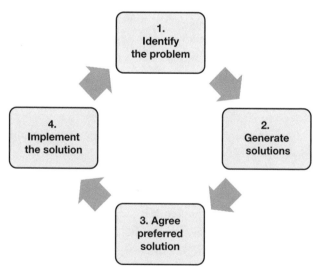

**Figure 1.1** Four-stage process for delivering relevant content

## 1 Identify the problem (or learning gap)

In order to find a solution, the problem will need to be correctly diagnosed. This is where you get to play the part of a detective! You will need to ask lots of questions. Could a customer service issue be about a service you are providing or is it the personal service received? Is it a lack of product knowledge or poor interpersonal skills?

## 2 Generate solutions

This is where you and your client's creativity takes place. Brainstorm as many solutions as you can without going into detail about whether or not the solutions will work at this stage. The aim is to end up with options that you can go through in more detail at the next stage. Create choices.

## 3 Agree preferred solution

Go through the list that you generated for stage 2 above. You may want to separate the solutions into specific areas, such as timeliness and cost. For example, it may be that a workshop is needed in three days. Therefore, a quick solution may be sought. Alternatively, there may be a very small budget, so that may limit the list of solutions you have.

## 4 Implement the solution

This is all about actions. The design, methodology and outcomes need to be defined and communicated. How will the training be delivered? What will success look like? How will it be assessed and measured?

# Questioning skills

There has to be a mix of open and closed questions to unpack all the information that you will need to run the workshop. You should determine the purpose and establish the client's objectives/goals, as well as your own.

There are so many factors involved when running a workshop that you have to be constantly one step ahead while being in the present. With each session that you are delivering, ask yourself, 'Is the audience currently engaged?' and 'Is the next intervention going to keep them engaged?'

## Closed questions

A closed question can be answered with either a single word or a short phrase. Therefore, 'How old are you?' and 'Where do you live?' are closed questions. Closed questions can also be used to receive a 'yes' or 'no' response.

### Using closed questions

Closed questions have the following characteristics:

- They give you *facts*.
- They are *easy* to answer.
- They are *quick* to answer.
- They keep control of the conversation with the *questioner*.

## Open questions

An open question is likely to receive a longer answer. Although any question can receive a long answer, open questions deliberately seek more information and are the opposite of closed questions – see the table opposite for some examples.

### Using open questions

Open questions have the following characteristics:

- They ask the respondent to *think* and *reflect*.
- They will give you *opinions* and *feelings*.
- They hand control of the conversation to the *respondent*.

| Question type | Purpose | Question form | Illustrations |
|---|---|---|---|
| OPEN | To establish rapport | Contact | Introductory questions and comments at the superficial level to put people at ease, e.g. at an interview or meeting. |
| | To show interest/encouragement | Non-verbal encouragement | 'Umm?' 'Ah?' Together with appropriate facial expressions. |
| | | Supportive statements | 'I see...' 'And then...?' |
| | | Key word repetition | Repetition of one or two words to encourage further response. |
| | | Mirror | Repeat back as a query. |
| | | The pause | Allied to various non-verbal signals. |
| | To seek further information | Simple, interrogative | 'Why?' 'Why not?' |
| | | Comparative | 'How do your responsibilities now compare with those in your last job?' |
| | | Extension and precision | 'How do you mean?' 'What makes you say that?' |
| | | Hypothetical | 'What would you do if...?' 'How would you feel if...?' |
| | To explore in detail particular opinions/attitudes | Opinion-seeking | 'How do you feel about...?' 'What do you think about...?' 'To what extent do you feel...?' 'Just how far do you think...?' |
| | | The reflection | 'You think that...?' 'It seems to you that...?' 'You feel that...?' |
| | To demonstrate understanding/clarify information already given | Summary | 'As I understand it...?' 'If I've got it right...?' 'So what you're saying is...?' |

Use link questions to move smoothly from one type of question to another, e.g. 'You mentioned just now that... How does this affect your department's work?'

## Counter-productive questions

Do not use counter-productive questions. The aim is to get the client talking, not to suggest 'right' answers; embarrass, confuse or mislead them; prevent them from saying anything; or discourage them – see below for some examples.

| Question type | Effect | Question form | Illustrations |
|---|---|---|---|
| COUNTER-PRODUCTIVE | To prompt desired answer | Leading | 'I take it you believe that...?'' 'You don't *really* think that...do you?' 'You must admit that...?' 'Isn't it a fact that...?' |
| | To confuse or mislead | Multiple and marathon | Two or more questions presented as a package. 'You did say you wouldn't mind being away from home occasionally? Oh, and you do have a current driving licence, don't you? I presume it's clean? And, err, by the way...?' etc. |
| | | Ambiguous | Asking a question in a rambling, incomprehensible way. |
| | To prevent respondent from saying anything | Rhetorical | Answering your own questions. 'Do you...? Of course you do.' 'I always say that...' etc. |
| | To discourage respondent/ indicate bias | Discriminatory | 'When do you intend to get rid of the older members of your team?' |

To have the ability to ask relevant questions is one of the most underrated skills. Practise your questioning skills on a colleague or a friend. Create a learning need scenario for their 'team' and try to ask the right questions to elicit the responses you require, so you can make the desired impact at the workshop.

## Types of questions

Below is a list of all types of questions that will be useful for you.

*Key questions to ask clients*

- Who is the target audience?
- What is the reason for this intervention?
- If there is an identified problem, how has this been addressed already? What were the results?
- Is what you need the same as what you want?
- Do others share your thinking?
- How receptive to training is the audience?
- What is in/out of your control in achieving your goal for the workshop?
- Is the style required one of a trainer or facilitator?
- What do you want people to be doing/saying at the end of the workshop?
- What will success look like to you?
- How will this success be measured?
- By attending this workshop, what will this enable you and the audience to achieve/do?
- Are there any alternatives to a workshop to achieve your goal – for example, a team meeting, buddying up, one-to-one conversations, etc.?
- On a scale of 1–10 (with 10 being high), how challenging should the facilitation style be? (This will depend on sensitivities and organisational culture.)

*Additional questions to ask an external client at the outset*
(These questions could also be relevant if you work for a larger organisation and the client is a department within your organisation that you have little information about.)

● What does your department do?

● What are its key responsibilities?

● How many employees are there?

● What are your biggest challenges, both internal and external?

● What are your key performance indicators (KPIs)?

● What is crucial to the success of your department?

**brilliant** tip

If you are being asked to provide training for a team of people, prepare a list of open questions that you need answered at a meeting or during a conversation. Keep adding questions as you see fit and use it as a template to give both you and the client confidence.

So, when you meet with a client, be prepared and organised. Use the training/facilitation planning template in Figure 1.2 to help you.

| **Facilitation plan for:** |
|---|
| [Brief description] |

**What is the overall objective?** Why are we doing this? What's the impact or benefit? What are the objectives of the change? What are the desired outcomes of this facilitation?

PRIMARY OBJECTIVES:
•

SECONDARY OBJECTIVES:
•

**Background.** Any relevant background information relating to the facilitation – current situation, symptoms, problems, performance gaps, etc.

**Scope.** For the facilitation and/or possible solutions, what does this brief include and not include? What is/is not important? Who is involved/not involved?

**Roles.** Who is the sponsor? The business lead? The facilitator? Participants? Subject matter experts?

**Constraints.** Are there any? What should be taken into account?

**Process.** What tools and techniques to use? What will be the overall approach to the facilitation? What style of facilitation is most appropriate?

**Group dynamics.** Is there anything that may help or hinder the facilitation? How do these people normally work together? Are there any particularly strong or quiet personalities? Are there any reactions to this facilitation that we should expect?

**Expectations.** What does the client lead expect of the facilitator? Is there anything specific that they want the facilitator to focus on?

**Deliverables.** As well as the objectives, what other deliverables are required (e.g. reports, manuals, etc.)?

**Pre-work.** What pre-work will be required?

**Environment and logistics.** Room layout? Equipment? Lunch? Breaks? Refreshments?

**Date(s):**

**Figure 1.2** Training/facilitation planning template

## Contracting

The model of 'contracting' in Figure 1.3 can be used at the planning stage, when finding out what a client really needs, as well as with a group of participants at the beginning of a workshop. At a workshop, it is essential the group answers all the questions.

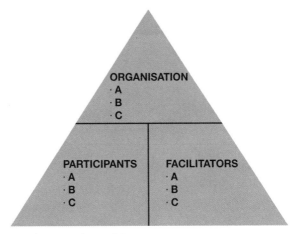

**Figure 1.3** The contract model

The 'contract' is a vital part of effective facilitation. It is not a legal document, but a clear, agreed statement of what is going to be achieved – usually for the three parties involved. By focusing on each area individually it can provide clarity around the expectations for each role. This is similar to the 'What's in it for me?' question. For each category, list three areas/actions that each party will commit to.

1  **The organisation/department/sponsor**
   List three benefits to the organisation if this session/ workshop is successful.

2  **What the participants expect to achieve for themselves and how they will work together**
   What will the participants commit to when working together to ensure success?

### 3  What we as facilitators need to do to ensure you get the most from the day/session

This is what the facilitator(s) will do to ensure success. This may be their style, their approach, trainer or facilitator role, etc.

---

## brilliant example

The contract was used at the beginning of a workshop on 'Improving Internal Customer Service'. The facilitator took the group through each section and got them to answer all questions.

### Benefits to the organisation

Question asked by facilitator – 'If we had a really successful day today, what would be the benefits to your organisation?'

● Reduction of 'silo' working.

● Staff talking more with each other.

● Reduction of internal email.

### What the participants expect to achieve for themselves and how they will work together

Question asked by facilitator – 'If we achieved our goals today, how do you, as a group, need to work together?'

● Lots of small group exercises.

● Lots of energy.

● Be honest and open with each other.

### What I, as your facilitator, need to do to ensure you get the most from today

● Show enthusiasm.

● Keep the tempo up.

● Share any top tips or knowledge with us.

---

 recap

- Never assume what someone wants is what they need.

- Know what particular style of facilitation you need to demonstrate and why – depending on the audience.

- Know the difference between facilitator and trainer.

- Be prepared to challenge assumptions.

- Ask lots of questions – play the detective.

- Use 'contracting' with the client and the participants.

# CHAPTER 2

# What can I do?

S uccessful workshops are dependent on effective facilitators. How effective are you? How do you know? Get to know yourself. In this chapter you will complete a self-assessment questionnaire to look at yourself through your own eyes, as well as those of others – such as the client and the audience. Learn to play to your strengths and know what these are, and be able to accommodate for your weaknesses. Are you required to be a trainer or facilitator? Knowing and focusing on what you can do, rather than on what you can't do, will assist in building self-confidence. Here we go... hold up that mirror and like what you see.

 Everything flows and everything is constantly changing. You cannot step twice into the same river, for other waters are constantly flowing in.

Heraclitus, 500BC

Always look to be even better than the last time. Situations change. Audiences change. The internal and external environments change. Therefore, you need to be adaptable to the changes around you.

## Hold up the mirror

Self-assessments are useful to see how you perceive yourself in the role of facilitator. If you work in a team, you could get feedback from others that will add value to your initial assessment.

So, as a start, list under each of the headings in the table below how you feel you contribute to being a good facilitator/trainer. Some examples are given to assist you.

| Your skills | Your abilities | Your attitudes |
| --- | --- | --- |
| Asking questions to generate ideas | Problem-solving | Positive |
| Non-verbal communication | Brainstorming | Neutral |
| Listening without interrupting | Ability to deal with conflict in groups | Enthusiastic |
| Enabling everyone to be involved | Understanding group dynamics | Respectful |

You will need to establish your preferred facilitation style – one that you are comfortable with. Obviously you should also be flexible about the range of audiences you may come across. Therefore, you will have to call upon your list of skills, abilities and attitudes, depending on the situation and the context.

There are areas that you will be comfortable with and you can also identify your areas of uncertainty.

Use the simple diagram shown in Figure 2.1 and some of the questions below to help you write down your self-perception on a sheet of A4.

- What skills, techniques and attitudes do I have? What do I need?
- Where are my strengths?
- Where am I comfortable in facilitation?
- What have I had little experience in or with?
- What are my initial worries, concerns or fears?
- Where am I uncomfortable in facilitation?

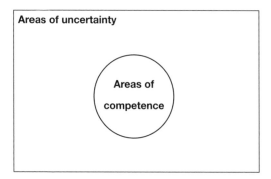

**Figure 2.1** Identifying your areas of competence and uncertainty

 **tip**

Remember, after you have planned everything, you need to choose your attitude. How you behave influences how others respond.

## Nine key ingredients of effective facilitation

1  Outline the objectives, structure and approach.

2  Keep in mind the difference between presenting and having a conversation.

3  Model effective behaviour.

4  Listen and show understanding.

5  Challenge those who do not build on dialogue.

6  Give everyone room to be involved.

7  Keep discussions focused and manage interruptions.

8  Summarise the main themes and insights.

9  Put questions to the group.

# Perceptual positioning

Perceptual positions are a very powerful aid to look at your own performance, especially when you might have some anxiety about situations. There are a number of perceptual positions but the following two will be effective.

● **First position – own perspective**
Our own personal point of view, our own reality, what we think and feel from our own experience.

● **Second position – other's perspective**
Imagine how you look, feel and sound to another person. The advantages of being able to use these positions are enormous; giving you insights and understanding while maintaining a positive personal state. It is the key to self-awareness and personal development.

## brilliant exercise

Take a blank sheet of paper and draw two lines to make four quadrants as shown on the next page. Answer the questions based on your facilitation/ training skills, identifying both strengths and development areas. You can adapt the quadrant headings as you see fit and also add further questions.

1  **How you see yourself**
This is your self-perception. Be honest with yourself. Do you have gaps in your skills or your knowledge? If so, what will you do about this?

2  **How the audience may see you**
Try a bit of visualisation. Place yourself in the audience – you are on the receiving end of one of your workshops. Now see yourself at the front of the room. What will be the first impressions? Is your voice shaky? Do you move around too much or are you static?

3  **How your client sees you**
Now see yourself through your client's eyes. Why are they using you?

Do they have to or is it by choice? What do they know about you? How might they perceive you and why? Ultimately, do they have confidence in you to do the job?

**4   How your peers or manager see you**
Aim to get feedback from those who know you best or have observed you running a workshop or session. How do they see your chances of success? Can they give you any advice from their experiences?

Crave the feedback. See feedback as a gift. Not as your enemy. It is your most valuable asset. Ask for it and receive it − and just say thank you!

| **1** How you see yourself | **2** How the audience may see you |
|---|---|
| Skilled/unskilled | Nervous/anxious |
| Confident/unsure | Confident/unsure |
| Credibility/no credibility | Professional/unprofessional |
| **3** How your client sees you | **4** How your peers or manager see you |
| Experienced/inexperienced | Strong/weak facilitator |
| Trainer or facilitator | Too soft/aggressive |
| What is their explanation for how they see you? | What is their explanation for how they see you? |
| Why did they choose you? | |

Place as much information as you can in each quadrant. You are only making assumptions at this stage. It may highlight areas where you and someone else may have different assumptions. For example, you may see yourself as confident but your audience may see you as anxious or nervous, for many reasons.

Self-assessment on its own is only half the story. The value is obtaining feedback from others and making your developmental decisions based on all the feedback. I observe many facilitators who think that they have not done a good job after the event. This could be that they think they struggled with team dynamics, getting the message across, keeping the audience motivated or failing to answer a question. This is their perception. When I have given them my feedback – my perception – it can be very different, and invariably is. This is why it is so valuable to receive as much feedback as you can. Get people to observe you in particular sessions and get them to look for certain areas that you want feedback on. Without feedback we just have our own self-perception. To be a brilliant facilitator, we cannot rely solely on self-perception.

## brilliant tip

If you want to be the best you can be, you can always place yourself outside your comfort zone and get yourself filmed while facilitating. It does take courage though. Watching it back really challenges your self-perception. My biggest learning was that I found myself saying 'um' in between sentences to fill the gaps. I would never have realised that unless I had witnessed it with my own ears and eyes. It has taught me how to use silence effectively. Knowing what you need to work on allows you to continuously improve.

In Mike Clayton's book in this series, *Brilliant Influence*, he states 'being likeable is a vital source of influence'. Positivity and negativity are contagious. Few people arrive at work to seek out the negative people and deliberately spend many hours in their company. A facilitator has to be positive and have the ability to influence those around them. People want to be around positive people. It is a great place to be. 'Can do' rather than 'can't do' attitude. If you reluctantly deliver workshops or you have little motivation to be there, then you should really think about your job. Having the energy and commitment to get up in front of others, in some testing times as well as enjoyable ones, requires this positive mindset. You will have a group of people in front of you. Some may want to be there, others may not. What are you going to demonstrate that will make the day worthwhile for them? Understand where you are, where you want to be and how you plan to get there.

> ### brilliant tip
>
> Each person in the audience will be asking the question, 'What's in it for me?' – the WIIFM factor. You will need to answer this question in your design and delivery.

## Rehearsing

### Why rehearse?

- Rehearsing your talk/session can reveal weaknesses in the design. It may be inconsistent, awkward in parts, overdeveloped in some areas and underdeveloped in others, or have poor transitions.

- Rehearsing your talk also allows you to identify and correct weaknesses in your delivery. You may, as a result, change the pace of the talk, improve your audibility or insert pauses. This is your real chance to test out your 'music' and 'dance'.

- It is important to check the duration of your talk. Allow plenty of time for questions. Decide whether to take questions during or at the end of each session.

- It is important that you rehearse with your visual aids. This will give you practice in using them. You can also reassess how useful they are; you may wish to re-design them.

### How to rehearse

- You should rehearse your session/workshop several times.

- Practise your talk aloud to check its clarity and duration – it will not be as telling if you run through it in your head.

Therefore, you might ask someone to listen to you, which has the advantage that you can get feedback. This may be helpful if the person has a similar level of knowledge of your chosen subject as you expect your audience to have.

- You might like to video a rehearsal and play it back. This can help build your confidence and help highlight areas to improve. Or use audiotape to hear how it sounds.

- Ideally, you should rehearse in the room that you will be presenting in.

- Memorise ideas not words – learning the words may destroy your spontaneity. Learn your storyline and sequence instead.

A useful metaphor for looking at your development needs is one based on the learner driver experience.

### Mirror

*Look* at where you are now and where you would like to be.
*Look* at your learning preferences and learning blockages.
*Look* at the effect your development may have on those close to you.

### Signal

*Tell* the appropriate people something about your development activity.

### Manoeuvre

*Choose* the best method, release the handbrake and move forward!

## brilliant recap

- Identify your strengths and development areas.

- Begin to see things through the eyes of others.

- Always ask for feedback, realising the value that it may not always be what you *want* to hear, but it will be what you *need* to hear.

- Practise asking and answering questions in a group setting. This will build up your confidence.

- Always rehearse – in everything you do.

# CHAPTER 3

# Planning and design

This chapter contains powerful templates that will enable you to draft your first outline and build on it. You will identify when detail is required, as well as any need for creativity and flexibility. We look at the pitfalls and what you should have in place to give you the confidence to design effectively, including timings and up-front evaluation. The wants and needs of the audience are the most important, and this chapter consolidates the learning of that in the previous three. Pre-work pros and cons are also addressed in an easy-to-follow list.

## brilliant tip

Do not take planning and design for granted. It is just as important as the 'doing it' part.

There are many project management books on planning. The message is the same. To fail to plan is to plan to fail. When I meet 'new' facilitators, most of their effort and energy is spent focusing on the part where they have to stand up in front of an audience, as this is generally their biggest challenge. They tell me that the planning and design is the easy bit and the fear is the doing. (Many experienced facilitators also experience this.) They tend to focus their energy on what they will say when that moment comes, when they may get tongue-tied or feel anxious or appear nervous. When I get them to review their design and planning of the workshop, it generally tells them that if they had spent more time on this, it would increase their confidence when asked to 'stand and deliver'. Alternatively, many facilitators get their buzz from the delivery, so the design part isn't as exciting for them. We will focus on how the design can be exciting as well. Either way, this chapter will give you tools and templates to assist you in designing your workshops effectively.

## Planning for success

When someone initially engages you to run a session or workshop, you must get your planning head on and be as positively challenging as you can. In Chapter 1, we said that many clients do not know what they need or require and your job is to ask as many questions as possible. Now that you have more of an idea, you have to put it into a deliverable product for your client.

Think about the type of presentation/facilitation/training that will be required. You will find some information in the tables below to assist you and to get you thinking. Focus on the type of delivery, the purpose, the audience and implications on preparation and delivery.

### A briefing

| What is the purpose? | Who are the audience? | Implications on preparation and delivery |
| --- | --- | --- |
| To give information. | May be either internal or external. | Do you already know your audience? |
| To motivate internal staff. | May be either a large or small audience. | Think about what they need to know. |
| To update the client team. | | You may not need to use visual aids. |
| To inform the client on latest state of play. | | You may not need to use prompt cards/notes. |
| To outline plans before an assignment starts. | | You may not rehearse if the briefing is very informal. |
| To report back after an assignment. | | You need to be clear on the facts. |

## A proposal

| What is the purpose? | Who are the audience? | Implications on preparation and delivery |
| --- | --- | --- |
| To persuade or sell. | May be an existing client. | You must research your audience. |
| To give information. | May be an external client. | Who are the key decision-makers? |
| To prompt action. | May be a potential client. | What is their agenda? |
| To demonstate teamwork. | May be an internal team. | What are the criteria for the decision? |
| To follow the client's agenda. | May be a small audience. | You must rehearse. |
| | | Appropriate use of visual aids. |
| | | Strong closing messages. |

## A seminar

| What is the purpose? | Who are the audience? | Implications on preparation and delivery |
| --- | --- | --- |
| To filter information. | May be either internal or external. | Audience research likely to be very generic (e.g. size, etc.). |
| To provide translation of technical subject matter. | Usually a large audience. | Think about what they are expecting to hear. |
| To prompt action. | | Link with other speakers or theme. |
| To promote awareness of your services. | | Any underlying agenda (e.g. to sell)? |
| To demonstrate expertise. | | You must rehearse. |
| To network. | | Will be a more technical agenda. |
| | | More formal approach. |

## Workshop planning templates

Once you have agreed the type of approach required, you need to start to plan in more detail. There are many workshop planning templates that you can use and I would recommend that you get to know what your preference is, yet be open to adapting rather than sticking to the same one rigidly; flexibility is key.

# Structuring your talk

Your structure must include a beginning, a middle and an end. Decide how conspicuous you want your structure to be. A clear structure is preferable, but it may be better to take a difficult audience along a chain of reasoning, showing them at the end where they have got to.

Try signposting. It can be blatant – 'I have three points to make, 1…, 2…, 3…'; 'In summary, my three points were…' Or it can be subtle – 'Now that we've dealt with… we can examine…'

Here are some areas to think about when planning your structure:

- Use the KISS technique – Keep It Short and Simple.
- Remember where the peaks of concentration occur:
  - in the first few minutes;
  - when they realise you are concluding and the end is near.
- Decide how to structure your ideas. If you have five points, of which number five is the strongest and number one is the weakest, start with number four (closest to five, which is your strongest point), then cover one, two and three, and then close with point number five (your strongest closing).
- Summarise at regular intervals.

If you are talking about a topic where there are two sides to the argument, you need to be aware of where you stand on the issue both professionally and personally, as well as where the audience stand.

Present one side of an argument when you know:

- most of the audience are in favour of your view;
- the audience are your subordinates who rely on you for guidance;
- the audience have to make an instant decision.

Present both sides of an argument when you know:

- most of the audience are opposed;
- the audience are your superiors who want the full picture;
- the counter-argument has or is going to be made known.

If you are presenting both sides, and you favour the 'pro' side, start with the 'cons' and answer them with the pros. This will help you to avoid destroying your own case. Put down the other side of the argument to put your point forward.

Always prepare the other side of the argument, even if you are not going to present it, so that you can answer any questions.

 **brilliant** tip

Always structure your talk. Plan it like a story (see Figure 3.1). Chunk it down into smaller sections – it can seem less daunting by focusing on smaller elements rather than the whole.

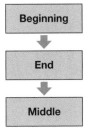

**Figure 3.1** The BEM model

To break down your design into smaller parts can be useful in many ways. Design may not be as exciting as the delivery for some people, but you can make it what you want it to be. When I design, I visualise myself in front of the room and the positive atmosphere that will surround me. This enables me to write in an excited and motivated mood. You can focus on key points in the session and make sure that each part is as important as the next. I always find it invaluable to break down any session that I am planning into a beginning, middle and end. The reason that the order has been changed in the BEM model is that the beginning and end are where you will have the most impact. It really helps to focus on these two sessions: first, the impact at the beginning where you can grab their attention, get buy-in, set the scene, expectations, etc.; and the end, ensuring that the objectives have been met and you have delivered a 'successful' workshop. Once these are in place, the middle part is the 'filling'. This should pull it all together. Just like making a sandwich, you cannot start with the filling; you need the bread on either side to make it!

## The beginning

- The beginning must be used to **grab the audience's attention**. They will be asking:
    - Who are you and why are you doing this?
    - What is the aim of the session/presentation?
    - Will it be interesting?
    - Will we be able to participate?
    - What will we get out of it?
- Create acceptance, build bridges with your audience, don't be offensive or unnecessarily outrageous, and get them on your side.
- Think about a general introduction:
    - relating yourself and your subject to the place and the audience.

- Think about a specific introduction:
  - make both your subject and your introductions clear.
- Your opening sentences might have one or more of these three aims:
  - to get the audience's attention and focus it on your presentation;
  - to give them a taste of what is to come;
  - to outline the content and structure of your presentation.
- You should learn your opening sentences. This reduces nervousness on the day.
- Having learnt your opening sentences your talk should start well. This will help establish your credibility and create a good rapport with your audience.
- Find out whether someone else will be introducing you. If this is so, it may be suitable to have ready a few words of thanks; if not, prepare a brief introduction for yourself. If you are presenting as part of a team, link your part with a previous talk: 'As Mike said earlier...'

## The end

- A clear conclusion is vital, linked to your original objective.
- Review what you have said with a summary of key points or an example which ties everything together.
- State or imply what you expect your audience to do, believe, know, etc., as a result of your session.
- Ensure you have met audience needs and have expressed any benefits in terms of benefits for them, not for you.
- **Never** introduce new points in your conclusion.
- Leave plenty of time for questions, if applicable.

## The middle

● The middle is the 'filling' – a key part of the session. You should aim to **keep** audience attention.

● Get all your ideas on paper first – let them flow in any order. Then categorise:

  – What are your audience needs?

  – What do you know about your subject?

● Choose three or four points or main categories from the subject notes that are of major interest to the audience.

● Select a logical structure to cover each main point, supported with the relevant subheadings.

---

### ↗ brilliant exercise

1  Write down in **one sentence** the message of your talk or presentation.

2  Refer to your brief. Read between the lines for a **hidden agenda**.

3  Identify the **purpose** of your talk/presentation. **Why** am I making this presentation? Is it to motivate, inform, persuade?

4  Note **three key messages** that you would like the audience to take away with them.

I always find that doing this exercise really gets me to focus on having a clear overview of what is required, and why.

---

### ▶ brilliant example

Here is an example of delivering an appraisal workshop for a client.

● **One sentence**: I am informing an audience of managers about the changes in the appraisal process and the impact these changes will have on them.

- **Hidden agenda**: On searching through the supporting documentation, there was some small print about performance-related pay linked to attendance at work. This was never brought up in the planning discussions even though there would be a high impact on managers and staff.

- **The purpose**: To engage managers in the process, as without them supporting the changes, appraisal will not be effective. To inform and influence the managers to take ownership of the process and to manage people effectively.

- **Three key messages**: Managers own the appraisal process, not HR; appraisal is not optional, it is an integral part of being a manager. Support the changes – they have been agreed and no amount of time challenging them will have any effect on the outcome. Use your precious time sensibly elsewhere.

## Designing your first draft

When designing your first draft, make a list of the key content or learning that *must* be covered. You will have discussed this with your client. Forget time limitations at this stage. If something is important you will always find the time for it. Do not worry about it now.

The following table can be useful to complete at the outset.

| Section title | Learning aim and outcomes | Methodology | Resources/materials |
|---|---|---|---|
| This column is the overall title you give to a particular session. | Enter here what the aim of the session is and what the audience will learn from it. | The type of intervention you have chosen to make the learning effective. | There may be just you or you may require other resources to assist you. You need to identify what/who they are. Do not forget the size of the room. |

### brilliant tip

If the participants do not know each other, you may want to run a short session where they are all introduced to one another. Do not fall into the trap of running this session at every workshop as it can be tedious if everyone is already known to each other. Ask your client if it is needed and if it will add value.

A simple way of quick introductions is to use the four Ws – that is, the four 'W' questions that you need the answers to.

## The four Ws

- Who are you? A name is all that is required.
- Where do you work? It is useful to know where people are working, either in an internal department or, if a public course, an external company.
- Why are you here? Get an insight into their motivations.
- What do you intend to learn? Find out their personal learning objectives. Get them to write these down. If you have planned correctly, the content should cover them already!

### brilliant tip

When you ask why someone is present, never accept 'because my manager told me to be here'. It immediately tells you that the person probably does not want to be here – and you may need to be aware of this in case of disruption through lack of motivation. Pre-empt this response by always informing the group that when someone responds to this question in that way, you will be asking for an alternative answer.

The table below is based on an introductory session for a workshop. Remember, no timings at this stage. Just get your thoughts around what is required. Think of each key section individually. Do not worry about patterns or which session follows the next as this can get in the way of your focus for each session.

| Section title | Learning aim and outcomes | Methodology | Resources/materials |
|---|---|---|---|
| Welcome and scene setting | Participants are very clear about why they are present and what is expected of them. | Facilitator welcomes participants and introduces the day, summarising the objectives, timetable and format. | Facilitator<br><br>Flipchart<br><br>Flipchart pens |
| Introductions – four Ws<br><br>Who are you?<br><br>Where do you work?<br><br>Why are you here?<br><br>What do you need to learn? | Participants to know who else is in the room and to share their learning objectives. | Participants get into pairs and introduce their partner to the group, replaying their four Ws. | Ensure the four W questions are on the flipchart. |

Complete this exercise for each separate session in your workshop. You can add rows and just complete for each session; or you can complete separate pages for each session, then put them together like a jigsaw puzzle when it comes to the next planning stage. Again, I cannot stress enough that you have to do what works for you.

You can find downloadable templates at **www.brilliant-workshops.co.uk**

## Up-front evaluation

**brilliant** definition

Evaluation

A systematic determination of merit, worth and significance of
something or someone using criteria against a set of standards.

When working with the client during the design stage, it is
important to have a column in your design template called
'evaluation' or 'success criteria', as mentioned in Chapter 1.

Generally, people talk about evaluation as something they will
do only at the end of the workshop. While this can also be useful,
don't leave it solely until the end before you discuss it. You may
have designed a fun, energetic session, but if you can't measure
its effectiveness then should you be doing it? How will you separate emotional evaluation from logical evaluation? What does
the client want? Some people may *feel* that it went well but, as a
facilitator, is that enough for you and your client? Is feeling good
about a session part of the success criteria?

**brilliant** tip

Evaluation begins at the design stage. Build it in to each session.
Know what success looks like at each stage.

To build up your credibility as a facilitator, you need to make
sure that each session can be measured by the client, in terms
of success. We will be looking at evaluation in more depth in
Chapter 11, but when at the design stage, have it at the forefront
of your mind. This should be agreed with the client – and if they

are paying or sponsoring the request, it is what is important to them that should be measured, not what is important to you. Add value and demonstrate your professionalism by adding your ideas to the process. See the table below taken from a coaching workshop as an example.

| Section title | Learning aim and outcomes | Methodology | Resources/ materials | Evaluation |
|---|---|---|---|---|
| Your current coaching competence | Each participant to be aware of their current level of coaching knowledge and understanding. | Individual completion of the coaching self-assessment questionnaire. To complete at beginning and repeat at end of workshop. | The coaching questionnaire (x12 copies). | Scores collated at the beginning of the workshop. Measure against scores at the end to see shift in improved knowledge and understanding. |

Some clients will not be interested in this amount of detail. However, you are ensuring that you are providing real value that can be measured. It enables you to feel more confident because you are working to specifics, rather than generalities. Get used to identifying the evaluation criteria up front with your client. They will thank you for it! So, always make sure that it's not just what you do, but also how you do it.

## The case for pre-work

Pre-work is something that needs careful thought. Do you set a group pre-work? Why do you feel you need to do so? How will it add to the value of the workshop? These are questions that need answering.

Listen to your client. They will inform you if participants will embrace it or ignore it. It can be a very bad experience if you have circulated pre-work only to find on the day that people didn't know why they had to do it, or they just didn't do it. You then need to manage the expectations of those individuals who did complete it.

Remember, when you get a group of people in a workshop, for whatever length of time, they are absent from their day job. In financial terms, that is costing the organisation money in terms of time away from the role as well as investment in the training itself. Therefore, it needs to be as effective as it can be.

Pre-work can really add value and save time on the day. Why get people into a room then hand out a process to read, when they could have done it in advance? The time away from their job needs to be used wisely. If it means that everyone will attain the same level of knowledge required for the session so you can move forward quickly, then it is also useful.

Any pre-work should be clearly communicated to the participants. They need to know expectations and consequences.

## brilliant example

One of my clients sent the following out to their participants for a change management workshop. 'Please read the following articles relating to change management. To ensure you get the most from this day you must read and understand them before the workshop. Failure to do so will mean that you will lack the knowledge of the other participants and you may be holding them back while you catch up. You will need to manage this.'

This is very clear and lets the participants decide. It leaves them in no doubt what needs to be done and the implications of not doing it.

 **tip**

Any pre-work that is given to participants should not be optional. It is there for a reason. The way you communicate it is the key.

## Pre-work pros and cons

| Pros | Cons |
|------|------|
| Can help to make sure everyone begins with the same level of knowledge. | Only a few people may complete it – 'too busy', etc. |
| Makes participants feel at ease with the content of the workshop. | Participants may not see the relevance or benefit of completing it. |
| Can save time on the day checking existing knowledge and understanding. | Can be too complicated or simple. |
| Can create interest and a buzz about the workshop. | Participants did not receive it 'in time'. |

# Types of learning interventions

In the next chapter we will begin to think about our participants – who they are, what they do and how our design will meet their needs. In our design, we need to cater for all types of individuals. For the vast majority of your workshops, you will not know the learning preferences of each of the participants. You can ask your client about how people learn within the organisational culture, but they will only have a perception about individuals.

You will need to understand a range of learning approaches and aids and know their uses and the dangers of not getting it right. We will go through a few of them here.

## Learning approaches

| Type | Uses | Be aware of |
|------|------|-------------|
| Presentations/lectures | • To get a lot of information across.<br>• To keep messages consistent.<br>• To enable facilitator to control the room/pace, etc. | • Can be too short/long.<br>• Can be delivered too fast/too slowly.<br>• Lack of buy-in. |
| Group discussions | • To stimulate thinking.<br>• To get everyone involved.<br>• To get buy-in/ understand barriers. | • The dominant participants.<br>• The quiet participants.<br>• Participants not feeling they can be 'honest'. |
| Group exercises | • Keeps energy levels up.<br>• Can be creative to engage everyone.<br>• Can be fun/enjoyable – make a 'dull' subject come to life. | • Can deviate from the real issue.<br>• Can be *too much* fun and the key message can be lost.<br>• Can be uninspiring. |
| Case studies | • To apply knowledge and understanding.<br>• To generate discussion.<br>• To view issue as a 'third party'. For example, it can be easier than being self-critical of own team/ organisation. | • Can take too long to complete.<br>• Can be too easy or too complicated.<br>• Can be irrelevant. |
| Role plays | • Keeps energy levels up.<br>• Can be creative to engage everyone.<br>• To practise newly acquired skills in a 'safe' environment. | • Participants can be fearful.<br>• Managing the feedback.<br>• Can be too unreal and therefore perceived as a false situation. |

## Supporting tools

| Type | Uses | Be aware of |
| --- | --- | --- |
| Flipchart | • Quick response/getting a point across.<br>• Very visual – place on walls.<br>• Gets everyone involved in small group work. | • Small writing/too large a group.<br>• Using up a lot of paper.<br>• Participants who are colourblind using multicoloured pens. |
| PowerPoint | • Can get key concepts/models across clearly.<br>• For larger groups – everyone can see it.<br>• Can make it visually stimulating. | • Too many slides.<br>• Reading everything that is on the slides.<br>• Can be boring. |
| Internet | • Can show 'live' information.<br>• Can show current/relevant YouTube clips, for example. | • Faulty internet connections.<br>• Hiding behind the internet for the whole session. |

 **brilliant** recap

- Design and planning are crucial to your success as a facilitator.

- Use templates to plan effectively.

- Build in up-front evaluation.

- Know the learning interventions available to you.

- Start with the key content and build the timings around it, rather than the other way round.

**CHAPTER 4**

# The audience

Know who the audience is. You need to make it your business. Find out as much as you can about them. How do they learn? This can be as a collective or individually. In this chapter you will learn how to identify your potential supporters and your saboteurs, especially if the stakes are high. You need to know what makes them tick and adapt your style accordingly.

> People will forget what you said, people will forget what you did, but they will never forget how you made them feel.
>
> Maya Angelou

In many ways, your skills, confidence and experience as a facilitator will account for a high percentage of your success. To make success more certain, knowing your audience can give you an added advantage.

Some people find it harder to deliver to people that they know, others find it the opposite. There are positives and negatives to each – they are personal feelings. A good facilitator should take the emotion out of the situation and focus on logical outcomes.

There are various types of learning intervention available to you. We all learn in different ways. Teams and organisations learn in different ways. This is where building rapport with the client is crucial. You need to ascertain what will and will not work and why.

For example, if you are in front of a group of financial account-
ants, understand the environment they work in, what is important
to them and how they approach their work, as well as the ways
things are done there. My experience has taught me that to run
a day full of extroverted activities, with role plays and high level
theory, will not be received as well as a day using practical exercises linked to theory and allowing for reflective thought.

Likewise, with a marketing team, it will generally not be as effective if you include lots of individual working and little in the way
of creativity and expression of ideas. Although this is not always
the case and we must not stereotype, really get to know your
audience, the environment they work in and know what will
work for them.

**brilliant** tip

Get to know the language and jargon that the audience use and
incorporate it into your design and delivery.

Think about the following.

## Who are they?

- Find out who will be there and how many will be there.
- Divide the audience into generic groups or broad
  categories. This will help in positioning your presentation
  and give you some idea of their level of knowledge and level
  of interest in your subject.
- It may not be possible to do more detailed analysis than
  this, particularly with a large audience. If it is possible,
  obtain more detailed information about each individual:
  - experience;
  - education;

- job description;
- attitudes;
- seniority;
- role in any organisational politics;
- work pressures and priorities;
- attitude to change;
- level of knowledge;
- sensitivity to subject, etc.

● Try to identify the decision-makers, and find out how they have reacted to speakers/presentations in the past.

## Why are they coming?

● Again, it may be possible to split your audience into broad groups:
  - those who are there to be entertained – 'the joy riders';
  - those for whom attendance is compulsory – 'the captive audience';
  - those who are interested in the subject – 'those who **want** to know';
  - those who need information – 'those who **need** to know'.
● If your talk has been successful in their terms, what will this mean, i.e. what is 'successful' to them?

## What do they expect?

● Think through their expectations.
● Find out what interests them.
● Identify how they will benefit from attending.
● Include information which is appropriate to their range of feelings – are they hostile, cynical, worried, enthusiastic?

- If you express your message in terms of what the audience want, they are more likely to listen and accept what you say.
- They will always retain the right to switch off if:
  - they are bored;
  - they don't get the information they expect or need;
  - they don't understand what is said;
  - they are distracted;
  - they are threatened or offended.

## Individual learning styles

As mentioned in Chapter 3, each person is an individual and therefore has an individual learning style. As this book is based on practical rather than theoretical content, David Kolb's theory of experiential learning will be very briefly explained now.

David Kolb believes that 'learning is the process whereby knowledge is created through the transformation of experience'.

Kolb's four-stage cycle shows how experience is translated through reflection into concepts, which in turn are used as guides for active experimentation and the choice of new experiences. The first stage, *concrete experience* (CE), is where the learner actively experiences an activity. The second stage, *reflective observation* (RO), is where the learner consciously reflects back on that experience. The third stage, *abstract conceptualisation* (AC), is where the learner attempts to conceptualise a theory or model of what is observed. The fourth stage, *active experimentation* (AE), is where the learner is trying to plan how to test a model or theory or plan for a forthcoming experience. For more information, see Kolb, D. A. (1984) *Experiential Learning: Experience as the Source of Learning and Development*, Englewood Cliffs, NJ: Prentice-Hall.

**Reflector style**
- Needs space to think and plan.
- Gets involved in detail and analysis.
- Likes to have space to observe and absorb.
- Ideal learning comes from reading and practice in risk-free environment.

**Activist style**
- Enjoys new experiences.
- Thrives on challenge.
- Likes to have a go.
- Secondments and courses based on experience are ideal for learning.

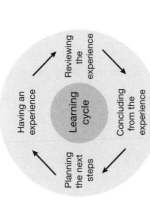

**Theorist style**
- Questions and probes the theory and methods.
- Likes to be intellectually stretched.
- Likes to be given space to analyse and reflect within a framework.
- Learns from models being explained and supported by experience.

**Pragmatist style**
- Links the subject matter to a problem to be solved.
- Likes the opportunity to practise practical things.
- Likes to do things immediately.
- Learns by solving problems in a 'real' environment.

**Figure 4.1** The learning cycle

*Source:* Based on data from Kolb (1976, 1981, 1984); Kolb and Fry (1975); and Honey and Mumford (1982)

Taking this work further, Honey and Mumford identified four learning styles which correspond to these stages. The styles highlight conditions under which learners learn better (see Figure 4.1). These styles are:

- Activist
- Reflector
- Theorist
- Pragmatist.

There are two versions of the Honey and Mumford Learning Styles Questionnaire – a 40-item and an 80-item version. More information can be found at **www.peterhoney.com**

By enabling participants to complete the questionnaire in advance, the facilitator can use the information to assist with design and to include exercises that meet everyone's needs.

### brilliant example

A colleague of mine sent out the learning questionnaire to participants well in advance of their workshop. This enabled him to design exercises for small groups that played to their learning preference. The majority were activists, so the exercises were based on the 'learning by doing' approach. Less talk, more action! The feedback afterwards was fantastic and they all thought that the workshop had been designed just for them!

Having the opportunity to discover learning styles before a workshop may not always be possible. There may be a shortage of time or it may not be totally relevant. You could always run this session at the beginning of the workshop to show how everyone learns differently.

- Activists prefer to **have** the learning experience.
- Reflectors prefer to **review** the learning experience.

- Theorists prefer to **conclude** from the learning experience.
- Pragmatists prefer to **plan the next steps** after the learning experience.

## Identifying the roles of resistance

If you are working internally for an organisation, you will have certain levels of access to the participants in advance of your workshop. This can be extremely useful in getting a feel for how the workshop is being perceived, who sees it as a need to develop or improve – and who sees it as interference to their day job.

Never assume that everyone who attends your workshops will want to be there. With experience, you can manage these emotions on the day, but some people may have already put the word round that it will be a complete waste of time and so the negative mindset has already been created.

The 'roles of resistance' is something that politicians use before canvassing for their votes. If you know the audience, or some of them, then you can complete this exercise and it will help you focus on who you need to engage with and who you shouldn't spend too much time with.

Think about the people attending your workshop and identify which of the five categories below they fall into.

| Who they are | What they do | Score |
|---|---|---|
| Evangelists | Champion of the workshop/content. Shouting it from the rooftops. | +10 |
| Supporters | Will support you. They are right behind you/the concept but will not lead. | +5 |
| Floating voters | Sitting on the fence. They could go either way. They need convincing. | 0 |
| Protesters | Against the workshop and will tell others when asked and can blow a lot of 'hot air'. | −5 |
| Saboteurs | Will deliberately set people off against each other. Actively against attending. | −10 |

## The evangelists

By identifying and knowing the people that are 100 per cent behind you/the content/training and development, you can rest assured that they need no convincing. They are your champions, your sponsors.

**Action**: There really is no need to spend time here as they are 100 per cent committed. Identify the participants that come within this category and score each one with a +10.

## The supporters

These people are also on your side. They may not be telling everyone or be as vociferous as the evangelists, but they will be fully behind the concept.

**Action**: All you need to do with supporters is to keep them informed about what is happening. This can be an email about the workshop or, if you pass them in the workplace, a positive conversation about it. Identify the participants that come within this category and score each one with a +5.

## The floating voters

These are generally the majority of the audience. They may not have strong feelings either way. However, you need to get to them before the saboteurs do!

**Action**: Spend most of your time here. You will need to engage and influence these participants. Go and talk to them and sell the benefits of attending and the learning outcomes to be gained. Take time to sit with them. It will be time well spent. Identify the participants that come within this category and score each one with a 0.

## The protesters

These people can talk a lot to colleagues about why they shouldn't attend and why the workshop will be a waste of time or boring. They talk a great deal but if challenged by a senior manager they won't show any resistance.

**Action**: Spend time here too. There is only a short distance to get them upwards to be a floating voter, so give them that time and sell the benefits. Identify the participants that come within this category and score each one with a $-5$.

## The saboteurs

You will always come across these people. No matter how great the design, the delivery and the feedback, they will never be convinced. They believe learning is something that others do and that they got where they are today by just doing the job. Their current knowledge and skills will carry them onwards and upwards. Of course, they may also not like the content or you!

**Action**: You do not need to spend any time here. It will be wasted. They will not change. To get them from here to a floating voter will take too long. Identify the participants that come within this category and score each one with a $-10$.

After initial identification, add up the scores and see how things currently stand. Although the scores are subjective, they can give you an overview of the types of challenge you may be facing when delivering. Think about who you will need to engage and influence in advance. Just talking to some of the floating voters can change their mindset about the workshop.

Again, knowing your audience in advance will help you manage expectations and build relationships, as well as influence those who need a little bit of extra understanding as to why the workshop is happening and the WIIFM factor – 'What's in it for me?'

 **recap**

- Get to know as much as possible in advance about your audience.

- Spend your time with people who are open to being convinced – floating voters.

- Become familiar with learning styles to maximise your delivery approach.

- Use the audience's language/jargon to explain concepts and theories.

- Always answer the WIIFM factor.

# CHAPTER 5

# Tools to support delivery

This chapter will run through icebreakers and energisers and will give you tips on how and when to use them. A quick checklist is provided to help you make your choice, as well as some exercises for all occasions that you can use immediately. There are a host of accredited tools that can add value, such as Facet 5/SDI®/Belbin/DiSC, which are acknowledged here from my personal experience and choice – but you will only get a reference for further reading and examples of how they may add value, as this book wants you to be able to deliver this week. Accreditation takes time and you need to decide which tool will work for you.

## Icebreakers

One definition of an 'icebreaker' is 'a beginning that relaxes a tense or formal atmosphere'. Do not fall into the trap of thinking that all workshops have to start with an icebreaker. They can be very effective if used for the right reasons, but to use them as a standard approach will not be received positively by everyone.

Designing an icebreaker and aligning it to the workshop is important. There are so many of them to be found in books and on the web that there is no shortage of options, as long as it is linked to the workshop topic, the new process or other outcomes. Icebreakers are generally fun and I recommend that the facilitator joins in too where possible – such as introducing a fun fact about each other. You are a big part of the success of the workshop. The participants will want to know a bit more about you. They can also be useful to get buy-in to the event.

Icebreakers can be used for a variety of reasons.

## Getting to know each other

- A newly formed team.
- An internal workshop in which you bring together people from different departments.
- A public workshop.

## Relieving any anxiety/tension

- A sensitive topic.
- Improving team communications.
- Managing conflict.

## Just for fun

- Starting on a high.
- Setting the tone for the workshop.
- The day is about fun or a reward for the team.

### brilliant tip

What is sometimes overlooked is the maturity of the group. No group is the same. Not everyone can or will want to run around outside or play tag or throw a tennis ball to each other to get them to reveal something about themselves. Have a few icebreakers up your sleeve to cater for diverse groups.

## Icebreakers you can use

Here are a handful of icebreakers that I have come across in my time as a facilitator. Some of them I have personally used, others have been recommended by my colleagues, and some have been used on me. I have no favourites – just remember to choose wisely to suit the occasion. If the group know each other, question whether an icebreaker is needed.

### Three facts game

Each person writes three things about themselves on a piece of paper. Two must be true and one a lie. Once completed, each person reads out their three 'facts' and the others have to guess the lie. This is fun and generally produces interesting information that others want to discuss.

**Useful for**: getting to know others.

### Three facts team game

A slightly different slant on the above, whereby teams get together and identify three things they all have in common. This should not be simple things like 'we all have hair', but should encourage the groups to really get to know each other and go deeper to find out fun things. Other teams again have to guess the two truths and the one lie.

**Useful for**: integrating new members to a team.

### Introduce your partner

Participants are split into pairs and each person has to introduce their partner based on what information you ask them to reveal. For example, three facts about themselves, three strengths, etc. This can help to develop listening skills and avoids people becoming self-conscious talking about themselves.

**Useful for**: getting to know others.

### A quiz

Preparing a quiz in advance is a good way of meeting many needs. Splitting the group into pairs or threes creates a sense of good-natured competition, and you can also add up the scores and offer a prize for the winner at the end. The questions could be about the organisation or about the subject you will be covering that day. They could also be about a particular process or policy. People can learn in a fun way, sharing the answers in the wider group and with you informing them what the correct answer is and why.

**Useful for**: increasing knowledge about a policy or process.

*Human bingo*

Create a card for each person with a grid of approximately nine boxes (three by three). Write generic themes in the boxes, such as the following:

● Has been to Asia.

● Owns a pet.

● Loves Indian food.

● Hasn't booked a holiday this year.

Then get people to go round and try to tick off as many of them as possible by getting someone who meets the criterion to sign their card.

**Useful for**: getting to know others.

If the group know each other but are unused to a workshop atmosphere, you can still get them to introduce another person for your benefit, as you will probably not know many of them. You could also run a quiz or get each person to talk about the last thing they learnt. This way there is no right answer so there is no pressure of getting it wrong in front of others.

# Energisers

There will be times when your audience loses attention. Some of this may be down to you and your style, but mainly it will be linked to forces outside your control. The first thing that you need to do is recognise when an energiser is needed. Negative body language, including yawning, and lack of responses to your questions can be clear prompts.

Reasons for losing attention:

- The room is too hot/cold.
- The topic is something they have already heard.
- They are not interested – they want to be elsewhere.
- Their minds are on other things.
- The speaker is not engaging.
- Peer pressure – colleagues in the room are losing interest.

Whatever the reason, you will need to inject some enthusiasm into the group, or simply call a time-out, so that thoughts and focus can be regained. Very simply, just by making people sit with different people or by moving seats can be very effective. It is OK to give people a five-minute break and it is also OK to let them know why you are doing it. Keep mixing up the pairs/ groups: sitting next to the same person all day, especially if you do not like them or they are negative, can be very wearing. Your role as a facilitator is to keep people on the move and learn from as many people in the room as they can.

## Five-minute energisers

Here are a couple of five-minute energisers that could be used.

### The tennis ball challenge

Get the group to stand in a circle and give them a tennis ball. They can throw it to a colleague only when they have called their name first. If it gets too easy, keep adding more tennis balls. It gets the blood pumping and you can also time them for a competitive edge.

*Birthday line-up*

Get the group to stand in a straight line in order of their birthdays from January to December. They cannot talk to each other. They have to arrange themselves in 'birthday' order using only non-verbal communication. Afterwards, the group can see how accurate they were. It can also show how communication can be received by others when no words are used.

## The impact of music

Another age-old question is whether playing music in parts of a workshop enables participants to focus better on their learning. It is common knowledge that music releases a chemical in the brain that has a key role in setting good moods.

In a 2010 study undertaken on behalf of 'Music Works For You' (a research body set up to demonstrate the powerful benefits of music to businesses and organisations, their customers and employees), research highlighted that '66 per cent of employees surveyed believed that background music made them feel better and more motivated at work, with over a quarter stating that they would be less likely to "take a sickie" if music was played at work'.

In their business research on the impact of music in the workplace, the findings were as follows:

- 87 per cent agree that playing music in the workplace improves morale.
- 86 per cent agree that playing music in the workplace puts employees in a happier mood.
- 69 per cent of workplaces recommend that others play music to improve productivity and staff morale.
- 62 per cent believe they can positively influence the behaviour of employees through playing music.

Music will create a huge range of emotions. Just think of the songs that make you want to dance, sing along or play air guitar. Songs that make you feel positive, that make you happy or sad, songs that make you feel on top of the world. We all have our own playlists. Famous music themes like those from the James Bond films or *Star Wars* evoke stirring emotions in many, while for me, when I was younger, the *Match of the Day* music made me want to go out and score lots of goals.

This is why some facilitators use music to create an environment where it is most conducive to learning. It can alter our state and increase our memory power. Many of you will know 'A, B, C, D', one of the most well-known children's songs to memorise the alphabet.

However, facilitators need to be aware that playing music can seriously backfire in a workshop.

### brilliant example

I did hear from a colleague that during an exercise he put on a piece of classical music that he thought would relax the minds of those present. After a few minutes one lady got up and ran out in tears. Apparently, it had been her wedding music and she had recently separated from her husband.

Some people play modern music, some play classical and some instrumental. It can have a range of effects on an audience as some may sing along and some songs may bring back sad memories that can have the opposite effect.

As with everything, we need to be very careful. There are pros and cons, just like all the other approaches. Try it out and see what happens.

## Even more tools...

The tools mentioned in the next few pages are just a handful available to you as a facilitator. If you work for an organisation they may have a preferred tool that they would like you to be using. As an external facilitator to an organisation, you really need to find out what tools the organisation is using and why, and whether it is in your interest to become accredited in any of them as there will obviously be a cost and time element.

Be very careful using tools to support learning as there have been many occasions when the tool has taken over and the *reason* for using the tool to support learning has been forgotten.

### Facet5

If you want to understand employee behaviours, motivations, attitudes and aspirations you need a psychometric profiling tool – and Facet5 is, to my mind, the best of the new generation of online profiling tools. It is based on a 'Big 5' factor personality model, which analyses people's will, energy, affection, control and – where this tool adds extra value – emotionality.

- Will – determination, confrontation, independence.
- Energy – vitality, sociability, adaptability.
- Affection – altruism, support, trust.
- Control – discipline, responsibility.
- Emotionality – anxiety, apprehension.

The system is very easy to use. Participants respond to an emailed invitation and are taken directly to the Facet5 questionnaire via the web. The data is processed instantly and is immediately available for review by authorised users, with reports shown on screen, which can be produced as a fully formatted 17-page PDF for printing or emailing. The report can also be compiled by obtaining 360°-observer feedback.

The report contains five different sections that can be used together to create a comprehensive overview, or individually to focus on key areas.

The areas are as follows:

1 **Overall profile**: Individual ratings against each of the above 5 facets.

2 **Family portrait**: There are 17 different Facet families and information is given against what an individual is like as a leader, what they are motivated by, their contribution to a team and what they are like to manage.

3 **Searchlight** (linked to six competencies): The competencies are leadership; communication; interpersonal; analysis and decision-making; initiative and effort; and planning and organising. It states what we should expect from the individual and what we should watch out for.

4 **Leading edge**: This section describes what their manager should do to get the best out of them. The areas are:
   - Creating a vision.
   - Stimulating the environment.
   - Treating people as individuals.
   - Goal setting.
   - Monitoring performance.
   - Providing feedback.
   - Developing careers.

5 **Overview of work preferences**: This section identifies core drivers and highlights which elements of a role will motivate or demotivate the individual.

*Uses*

● **Personal development**: Finding out about yourself in terms of how you act in certain circumstances and how what you see as your strengths can be seen as something totally different by someone else. Good for one-to-one coaching.

● **Team development**: A separate report called Teamscape identifies team roles and preferences. This is great fun and the many interactive exercises that support learning make it a memorable learning experience. Good for teambuilding and introducing new members to a team.

Website for further information – **www.icebergtools.com**

## Strength Deployment Inventory (SDI)®

The SDI is a proven, memorable tool for improving team effectiveness and reducing the costs of conflict. It is the flagship learning resource of a suite of development tools based on Relationship Awareness — a learning model for effectively and accurately understanding the motive behind behaviour. When people discover the unique motivation of themselves and others, they greatly enhance their ability to communicate more effectively and handle personal and interpersonal conflict more productively.

The SDI is not just another 'personality test'. It is a self-scoring motivational assessment tool that provides an understanding of what drives you and what drives others – an understanding that empowers you to communicate in a way that achieves the results you desire. The SDI reveals the responses to 20 questions. Of these, 10 are answered when you are feeling at your best and the second group of 10 are answered when things are not going so well. It uses a graphical score charting method, and individuals and/or teams can see everyone's results on a single illustrative triangle.

*Uses*

Another good tool to look at preferences, but this is particularly useful for looking at conflict and how it can be resolved. It is best used in teams, if there is either an issue or simply a new team getting together. There are good active exercises to understand your preferences and a wide range of facilitator resources are available to embed the learning.

Website for further information – **www.uk.personalstrengths.com**

## Belbin team roles

In the 1970s, Dr Meredith Belbin and his research team set about observing teams, with a view to finding out where and how differences in team performance come about. As the research progressed, it revealed that the difference between success and failure for a team was not dependent on factors such as intellect, but more on behaviour. The research team began to identify separate clusters of behaviour, each of which formed distinct team contributions or 'team roles'.

To find out which of the nine Belbin team roles you have an affinity towards, and which ones you do not, you need to start by completing a Belbin Self-Perception Inventory. This is a questionnaire that takes about 20 minutes to complete.

By completing your self-perception inventory you receive a four-page report that gives you an overview of your behaviour, along with advice on how to project your strengths and be aware of potential weaknesses.

The report can also be compiled by obtaining 360°-observer feedback.

## The nine team roles

| Role | Team contribution | Allowable weakness |
| --- | --- | --- |
| Plant | Creative, imaginative, unorthodox. Solves difficult problems. | Ignores incidentals. Too preoccupied to communicate effectively. |
| Resource Investigator | Extrovert, enthusiastic, communicative. Explores opportunities. Develops contacts. | Over-optimistic. Loses interest once initial enthusiasm has passed. |
| Co-ordinator | Mature, confident, a good chairperson. Clarifies goals, promotes decision-making, delegates well. | Can be seen as manipulative. Offloads personal work. |
| Shaper | Challenging, dynamic, thrives on pressure. Has the drive and courage to overcome obstacles. | Prone to provocation. Offends people's feelings. |
| Monitor Evaluator | Sober, strategic and discerning. Sees all options. Judges accurately. | Lacks drive and ability to inspire others. |
| Teamworker | Cooperative, mild, perceptive and diplomatic. Listens, builds, averts friction. | Indecisive in crunch situations. |
| Implementer | Disciplined, reliable, conservative and efficient. Turns ideas into practical actions. | Somewhat inflexible. Slow to respond to new possibilities. |
| Completer Finisher | Painstaking, conscientious, anxious. Searches out errors and omissions. Polishes and perfects. | Inclined to worry unduly. Reluctant to delegate. |
| Specialist | Single-minded, self-starting, dedicated. Provides knowledge and skills in rare supply. | Contributes on only a narrow front. Dwells on technicalities. |

*Source*: Based on data from **www.belbin.com**

## Uses

Good for teambuilding, new teams or new members. The report is simple and easy to follow. Have fun letting others guess their colleagues' role before they see the report. Their perception is their truth so it is important for individuals to know the impact of what they do and how they do it.

Website for further information – **www.belbin.com**

## DiSC

The modern DiSC theory first appeared in the 1920s, in William Moulton Marston's book, *The Emotions of Normal People*. Marston, who is probably best known today for his invention of the polygraph or 'lie detector', developed DiSC to help demonstrate his ideas of human motivation. Its simple 24-item questionnaire was intended to help him quickly appraise different behavioural types for analysis and comparison (Marston, 1999).

DiSC profiles are presented in a simple graphical form, which can be easily understood with only a little experience. Each question asks a candidate to identify which of four options is most applicable to them, and which is least applicable.

A DiSC test produces not just one, but three, distinct profile shapes: the Internal, the External and the Summary. These are based on analyses of different sets of answers, and each describes a different aspect of a person's behaviour. Each profile shape will come to the fore in a certain type of situation.

- The **Internal** profile reflects the person's true motivations and desires. This is the type of behaviour that often appears when an individual is placed under pressure.

- The **External** profile describes a person's perceptions of the type of behaviour they should ideally project. This shape usually represents the type of behaviour that an individual will typically adopt at work.

- In reality, people will usually act in ways consistent with elements from both the other types. The **Summary** profile is a combination of the other two profiles, describing a person's likely normal behaviour.

A DiSC test measures four main traits, or 'factors', of behaviour, from which the system takes its name. These are:

- **D**ominance – relating to control, power and assertiveness.
- **I**nfluence – relating to social situations and communication.
- **S**teadiness – relating to patience, persistence and thoughtfulness.
- **C**ompliance – relating to structure and organisation.

*Uses*

Good for one-to-one coaching – providing a high stream of data, including an individual's communication style, their value to the organisation, how to communicate with them and how they like to be managed and motivated. Teams benefit from this approach, with perceptions again to the fore. A detailed supporting workbook provides more knowledge and ensures learning is not left in the classroom.

Website for further information – **www.thetrustedadviser.com**

 **brilliant** recap

- Use icebreakers and energisers only if you have to. Do not automatically build them in and allocate time to them unless you have agreed that it is essential to the success of the workshop. Have a few up your sleeve and be ready to use them as and when required.

- Remember that the above tools are my personal choice. I have used them all and have had many fantastic days introducing them to teams and having fun while learning. There are so many others on the market and you will have to decide what works best for you and your clients. Of course, you do not have to use any of them.

# CHAPTER 6

# Handouts, workbooks or other resources?

n this chapter we will look at materials to aid learning. The participants want to walk away with something that will trigger positive memories of your workshop. We will consider examples and the reasons for printing everything or nothing at all. We will also look at other ideas and suggestions to provide participants with souvenirs, such as USB memory sticks, mouse-mats, etc., and the need to separate a gimmick from a useful learning resource. Not having any supporting materials during the course of a workshop will hinder learning ability, as most people forget what is said quite quickly, but they will remember how they approached something or worked through a problem-solving model, for example.

## Do you need supporting resources?

If you have attended a workshop you will probably have walked away with some resources that have aided your learning throughout the day and that are there for you to read when called upon in the future. I remember attending a public sector leadership event and we all walked away with a marvellously colourful thick spiral-bound book of unmemorable exercises and a very long list of further reading. It was very heavy and large and I don't know where it is now – and I dread to think of the cost of putting it all together. Content is king. Relevance is king. This certainly did not have the desired effect on me and many of my colleagues.

If you are using supporting resources, first focus on making them useful, not showy.

I have asked my colleagues about their thoughts and preferences on this subject and there really is no one right answer. However, you need to be mindful of the alternatives and the impact of using what you think is the right format. It was certainly agreed that supporting resources are essential for participants to get the most from a workshop.

A useful tool to support you when thinking about your approach is to focus on the following three factors:

### Time vs Quality vs Cost

When making a decision to do or consider something, the TQC model is very useful (see Figure 6.1). This is generally used in projects but can be useful to follow in a variety of situations around decision-making.

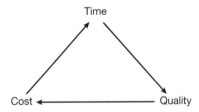

**Figure 6.1** The TQC model

*Source*: Adapted from the project triangle in Kerzner (2009). Reprinted with permission

## Time

- I need lots printed and they need it today!
- Speed or attention to detail?

- How does the quality or the cost impact on my decision?

## Quality

- What matters is that it's done properly.
- It's quality that matters.
- How does the time or the cost impact on my decision?

## Cost

- I want it done as cheaply as possible/I want no expense spared.
- What is the budget?
- How does the time or the quality impact on my decision?

Time, quality and cost all go hand-in-hand, but the rule of thumb is that you can only deliver two of the three. While it may seem harsh, it does make sense that a quickly and cheaply made product will lack quality refinements, for example. There will be exceptions to the rule but the principle itself has withstood the test of time. You will need to weigh up the importance of each of the areas. Do not let the need to balance these three core elements evenly concern you too much. They do have an impact on each other, just like the 'water-bed effect', whereby if you place pressure on (or sit on) one corner, then the other corners will rise up. Therefore, too much focus on one area will impact on the others. Aim to break down each area and analyse the impact all three may have – on each other and on the end product.

After all, you can't expect a cheap service to go along with speed and quality, or speed to go with cheap and quality, or even quality to go with cheap and speedy. Plan, plan and plan again.

 Nothing is more difficult, and therefore more precious, than being able to decide.

Napoleon Bonaparte

## Printing

Before we move into what could be printed to support participants, let us look at the advantages and disadvantages of colour vs black and white printing.

### Colour – possible perceptions of full colour documents/workbooks

| Advantages | Disadvantages |
| --- | --- |
| Initial high impact on reader. | Can be perceived as a waste of money. |
| Can bring documents 'to life'. | Extra costs associated – printer ink, etc. |
| Can show expense is not an issue – we are important as delegates! | May have colourblind/partially sighted participants. |
| Can take the attention away from a poor facilitator. | Can take the attention away from a good facilitator. |
| Can stimulate the senses. | Just like all the other handouts – nothing 'special'. |

### Black and white – possible perceptions

| Advantages | Disadvantages |
| --- | --- |
| Cheaper than colour – printer ink, etc. | Can make a long document/workbook seem 'lifeless'. |
| No colourblindness concerns with reading. | Participants may compare it unfavourably with a colourful document from a previous workshop. |
| Photocopying is faster than colour. | Seen as 'unprofessional'. |

There are many other considerations before you decide *how* to print. You need to know *why* you are printing something. Who is paying for the printing – you or the client? (I often send a PDF to the client and they print it to save costs.) Is it a supporting workbook or individual handouts? Both are valid and very good at cementing learning or emphasising a point.

## Questions to ask yourself

● Should I use single- or double-sided? What is the driver here? Cost-effectiveness or eco-friendliness?

● What type of paper should I use? Is it recyclable? Does it have to be? Quality sheet paper or budget?

● What font size and type should I use? Organisations will generally have a 'house' style. I personally work to Arial size 12 font unless I know in advance that someone who is attending is long-sighted and I have been informed to use size 14 or 16, depending on the individual. If you are not sure – ask!

**brilliant** tip

If in doubt, and you can be flexible, have the cover in colour only for initial maximum effect.

# Workbooks vs handouts

For clarity, I will refer to workbooks as being over 10 pages in length, containing theory and exercises that are completed as the workshop progresses. It is a participant's personal journal and will be a reference to the day. They are given out once.

Handouts are separate sheets (can be more than one page) that are given out at certain points in the workshop, to accentuate a point, support a theory or complete an exercise. These are my definitions.

As we have already said, ask the client what the audience prefer and what has worked in the past. Again, add value by challenging to see if a different approach will be effective.

To begin with, one approach is not better than the other. They are situational.

## Workbooks

I have found that it is so much easier to collate and print workbooks as you only give them out the once. Participants can then follow a structured approach through the day and know they have a resource to refer to in future.

| Advantages | Disadvantages |
| --- | --- |
| Easy to collate. | Participants can read ahead so future impact may be lessened. |
| Printed and bound can look professional. | Can feel like 'being back at school'. |
| Structured approach for all – as a facilitator you know what's coming next. | Once printed, you cannot change/ omit the exercises – lack flexibility. |

## Handouts

You will need to collate and place handouts in order of the exercises planned. There can be many of them so they need to be clearly numbered. If you are branding them with a logo, etc., then you will need to do each one separately – a workbook just requires you to enter them once in the header or footer.

| Advantages | Disadvantages |
| --- | --- |
| Can use as and when the need arises or not use at all. | Quantity may be confusing for facilitator. |
| Gives flexibility to facilitator to change structure of the sessions. | Not enough structure for facilitators who prefer to follow a process. |
| Can hand out when you think it will have the most impact – more control for the facilitator. | Unless you provide a folder, handouts can get mixed up/misplaced. |

**brilliant** tip

Always work closely with your client to share the benefits and disadvantages of supporting materials. Be sure that you balance the overview vs detail conundrum. It is easy to be criticised for providing too much or too little information. Get it right at the outset – together.

## Is it a useful learning resource or a gimmick?

There are two definitions, or two ways to look at what a gimmick is. A gimmick, as stated in the Oxford English Dictionary, is 'a trick or device intended to attract attention, publicity, or trade'. It can also be seen as 'an ingenious or novel device, scheme, or strategy, especially one designed to attract attention or increase appeal'. This second definition could also be seen as a resource.

We need to be aware of what we are doing to enhance our workshop or learning experience with our participants, as their perception will come under one of the two previous definitions.

Gimmicks are an essential marketing tool. Whether it is adding a feature to your training courses not found in competing products, or adding extra incentives to get organisations to choose you instead of the other company, these are all gimmicks. They can be both ingenious and devious. Whichever way you look at it, gimmicks are some sort of hook that will aim to get others to remember you (or your workshop) above all the others. It can also be a positive reminder to participants of the key learning points that were conveyed at your workshop.

We all want participants to leave our sessions saying that the day made a real difference to them and that they can apply their learning in a practical way within their workplace. Our initial way of doing this is to ensure that all the learning objectives stated at the beginning of the day have been met by the end of the day. When our participants leave, how do we help them remember these key things? Once they leave that room, there is no way of making people act on the things they have learnt. We can offer them the most compelling case to implement any actions they have agreed to our face, but how can we place some sort of learning trigger that will make it easier for them?

Obviously, organisations can ensure implementation by involving line managers and using all levels of evaluating the learning. (Evaluation is covered in Chapter 11.) They can walk away with our workbooks and our handouts but we cannot be sure they won't be next week's doorstops!

## Learning nudges

Therefore we need to think about little learning nudges that will give the participant and their manager every chance to make a difference to the organisation. The next few nudges may not work for everyone, but for some people they will be very powerful.

### Pens

Handing out pens is not going to encourage learning and committing to action. They are a marketing tool for the organisation or facilitator delivering the workshop. There is nothing wrong with this, as I can't remember a workshop where at least one person hasn't asked for a pen! They are a practical resource.

### USB memory sticks

These memory sticks can be given out at the end of a workshop and could contain your PowerPoint slides, reading material, links to relevant internet sites, etc. They can be branded with your company's logo or the title of your workshop. People do tend to remember better when there is a trigger such as this. If you wish, be prepared to make it 'read only' and protect it so that the information cannot be overwritten, otherwise participants may be tempted just to use it as a new memory stick for their own use (although this may be OK with you). Also consider putting your company marketing information on there . Be aware that this may leave the cynics complaining that it is just a branding exercise and that the memory stick is just a gimmick. Of course it is!

## Things to seriously think about

There are many other supporting resources people use to try to make the learning experience more enjoyable. Please proceed with caution. Get it wrong and face the consequences. Read on...

## Props

**brilliant example**

A colleague recently shared a story with me when they totally misjudged the use of placing stress balls and soft toys in the room. He had just completed his Certificate in Training Practice (CTP) qualification and was keen to try things out. He got to the venue early and laid them all out on the table. He said he was just so keen to get going that he forgot that he was delivering to an audience of finance professionals.

They came in and cleared them up and handed them back to him. He said nothing and left it at that! We laugh now and he still wonders whether they gave them back because they thought they had been left there by a previous facilitator, or whether they had given him a subtle nudge saying, 'Don't you dare...'

You can purchase 'fiddle kits' for participants that contain toys, motivational cards, bendy people and squeezy pencils to name a few. I have seen clockwork ladybirds, desk bells, finger puppets and hairy hedgehogs. Be careful with all these props – as with all approaches there will be supporters and cynics. Everyone learns differently, as we know.

Never let the assorted props take away from either the focus of the workshop or your personal impact. Use them wisely and ensure that they fit the occasion. If you are facilitating a group that requires lots of ideas and creative energy, then it would be more appropriate to use props than to place them in front of a group making decisions about detailed processes.

## Supplying snacks and freebies

The majority of facilitators/trainers that I know always recommend putting out either sweets or mints to enhance the atmosphere they are trying to create. Having a variety of snacks available can give people a sugar-rush for energy, for example. It can also relax people when they arrive as people share them out and begin conversations.

**brilliant** tip

To use the sweets or snacks in the workshop, get two of each type of sweet, for example, and hand them out to participants as they arrive. Then get them to pair up with the other person who has the same sweet and get them talking about the workshop.

Mousemats and mugs have been a common theme in workshops for a while. They can be used solely as a gimmick, with details of the facilitator's organisation on it, or they can contain useful information, such as:

- an organisation's values;
- a key step-by-step process;
- the key aims of the workshop;
- an inspirational quote.

 **recap**

Whatever you decide to do, from all of the above to nothing at all, remember it should be a useful learning resource that will benefit the participants and the organisation. Look around and see what is on people's desks. Ask what has worked before and what hasn't. Ask yourself, is it essential or desirable?

# CHAPTER 7

# Presentation styles and your 'stage'

Once the design is in place, we need to bring it to life. You will be in front of an audience. We looked at finding out about them in Chapter 4 and now you have to think about your style(s) of delivery. In this chapter we will look at a range of styles – how they match certain audiences, as well as encouraging you to be more aware of your style when you deliver by knowing your surroundings. The subject matter needs to be aligned to your style. Among others, do you use PowerPoint or flipcharts? We will take some time to look at your stage – the types of room layout available to you, placement of flipchart stands/projector, etc., and how you can bring a 'dull' room to life.

 If you don't have confidence, you'll always find a way not to win.

Carl Lewis

If you are starting out in a training team or department within an organisation, your level of seniority should not matter if you are going to facilitate a group effectively, although you need to have confidence in what you are doing. For many people, a big barrier is that the participants are more senior than them. You are not there to do their job or demonstrate you are better at doing their job than they are. Do not worry yourself about this. You are there as the enabler. You are there to ensure that they get the most out of the session and that the environment supports what they want to achieve. You are demonstrating a very different skill set.

## Preparation

There is a saying that a successful presentation is 90 per cent preparation. As a facilitator you need to make the time to plan and prepare.

### Identify your objectives

Make sure that you know why you are in front of a group of people and what you are there to achieve. This may take a long time to get pinned down but it is essential you are 100 per cent clear about your role. Once you get this right, the other elements of preparation can begin.

### Research your audience

As was covered in more depth in Chapter 4, learn all you can about your audience, from a variety of sources, and identify their objectives.

### Choose a logical structure – beginning/middle/end

Get the story right. It helps with confidence and can signpost your way through the session.

### Keep the content simple

Don't fall into the trap of just showing the audience how clever you are. By all means, if it is relevant to the subject then that is OK. However, please do not use it as an opportunity to showcase everything you have done and everything you know throughout the session. It can be quite wearing if it becomes the 'you show'.

### Think of real examples

To explain things that the audience can put into context, think of a few real examples, a few stories – preferably in their world. This will show that you have understood their environment and can give you credibility.

## Anticipate questions

Make a list of all the questions the audience may ask. Have the answers ready. This builds your confidence and can help you in the design stage if it uncovers an area you hadn't thought about previously.

## Design visual aids

These are covered in more depth later on in this chapter. Match the visual aid to the occasion.

## Rehearse aloud

All actors rehearse aloud. Be comfortable hearing your voice and be aware of where you place your emphasis in sentences and words. Know your pace and pitch.

## Learn opening sentences

Get your opening rehearsed and planned. It gives you confidence and sets the tone for the rest of the session.

## Check room layout

Although this is covered in more detail later in this chapter, ideally you should be in control of how the room is set up. This will help you plan all the other things, such as where you stand, flipchart placement, etc.

---

### brilliant impact

1  You should be able to write down in **one sentence** the message of your talk.

2  If you cannot write down in one sentence what the key message is, then you should refer to the brief you were given and see if there are any grey areas or multiple objectives. This one sentence will give you clarity and confidence.

▶

**3**  Refer to your brief. Try to read between the lines for a **hidden agenda**.

Is there a political or personal agenda sitting beneath all this? What
is the real reason for this workshop? Is it perceived as a reward or a
punishment?

**4**  Identify the **purpose** of your talk. **Why** am I making this presentation/
facilitating this team? Is it to motivate, inform, persuade?

This is very important as it will help you to focus on your delivery style.
To motivate others you need to be enthusiastic yourself. To inform others
you need to know your subject. To persuade others you need a good
argument.

**5**  What are the **three key messages** that you would like the audience to
take away with them?

Similar to the first point, to give yourself that focus that you are
pitching it correctly, state three key messages and then clarify them
with the client or a colleague. You can then evaluate whether the
session was successful against these criteria.

## The words and getting them right

When designing your workshop or session, you will obviously
be thinking about the words you will use to get across your key
messages.

### What you say

- Identify what information your participants need for you to
  convey your message.
- Avoid giving too many facts. Your participants are only
  likely to remember a few.
- Do not try to cover too much in the time you have been
  given.
- Adapt the talk to your participants, so that you are
  addressing them personally.

- When you are describing a new concept, try to liken it to something the participants are familiar with. This may mean using personal anecdotes or industry-based examples.

- Use facts to support your points, as they are indisputable, whereas opinions are not.

- Avoid using too many long words and jargon.

- Use simple sentences, as they are easier to listen to.

- Show respect and affection for your audience. Use the 'we' and 'our' form when you can.

## Focusing on key messages

- What **must** people know? These are your key points.

- What **should** people know? These points will help you to get your message across and help the participants understand.

- What **could** people know? These points are not vital for your workshop, but may make it more interesting (e.g. stories, examples, anecdotes).

# Creating the right impression

In terms of how you actually deliver the content of your presentation, you should remember the following:

- **Earn** respect by knowing your subject.

- Be **eager** to share your knowledge with others.

- Use **energy** to enhance what you say and this will help your body language to give the right messages.

Three key areas of a presentation can be summarised as:

- the music – the use of the voice;
- the dance – appearance and body language;
- the words – what you say.

Early research was done by Professor Albert Mehrabian into non-verbal communication and the impact that words, body language and tone of voice have on another person (Mehrabian, 1971). The results were:

- 38 per cent happens through voice tone (music);
- 55 per cent happens through body language (dance);
- 7 per cent happens through spoken words.

## The studies

Mehrabian and his colleagues were seeking to understand the relative impact of facial expressions and spoken words.

### Study 1

Subjects were asked to listen to a recording of a female saying the single word 'maybe' in three tones of voice to convey liking, neutrality and disliking. The subjects were then shown photos of female faces with the same three emotions and were asked to guess the emotions in the recorded voices, the photos and both in combination.

**Result:** The photos got more accurate responses than the voice, by a ratio of 3:2.

### Study 2

Subjects listened to nine recorded words, three conveying liking ('honey', 'dear' and 'thanks'), three conveying neutrality ('maybe', 'really' and 'oh') and three conveying disliking ('don't', 'brute' and 'terrible'). The words were spoken with different tonalities and subjects were asked to guess the emotions behind the words as spoken.

**Result:** Tone carried more meaning than the individual words themselves.

## The misunderstanding

Mehrabian's resultant formula is generalised to mean that in *all* communications:

- 7 per cent happens through spoken words;
- 38 per cent happens through voice tone;
- 55 per cent happens through general body language.

Of course this cannot be true: does an email convey only 7 per cent? Can you watch a person speaking in a foreign language and understand 93 per cent?

## The implications

While the exact numbers may be challenged, the important points can easily be lost in the debate about how valid or not the study was. Useful extensions to this understanding are as follows:

- It's not just words – a lot of communication comes through non-verbal communication.
- Without seeing and hearing non-verbals, it is easier to misunderstand the words (telephone/email, etc.).
- When we are unsure about words and when we trust the other person less, we pay more attention to what we hear and see, rather than the words spoken. So, where the words did not match the facial expression, specifically in Mehrabian's research, people tended to believe the expression they saw, not the words spoken.

## So what?

- Beware of words-only communications like email. It is very easy to misunderstand what is said, even if emoticons (smileys) are used.

- When you feel that a person is not telling the truth, check out the alignment between words, voice and body.

## The music

Your voice is the most effective way of:

- bringing words to life;
- making people listen;
- making what you say interesting.

**brilliant tips**

- Speak conversationally to come across as both warm and friendly – but don't lose volume.
- Read in a dramatic way – be 'over the top'. You must sometimes overact just to sound normal.
- Do not stifle your emotions – you will just sound *too* under control.
- Do not be frightened of pauses. Pauses give structure and make what you are saying easier to understand.
- Speaking more quietly than normal can have just as much impact as speaking loudly.
- Use intonation to stress key words.
- Talk to, not down to, your audience.

You need to use your voice to your full advantage. Get used to hearing it in large groups. Volunteer at meetings to have a go.

Complete the table on the next page. The desired column is what you are striving to demonstrate. The undesirable column is the opposite. Tick if you think you sit in column 1 or 2. For any ticks in column 2, note what options are open to you to move across to your desired outcome.

## Your speech profile – checklist

| Desired | Undesirable | Column 1 or 2? | What do I need to do about it? |
|---|---|---|---|
| Enthusiastic | Lifeless | | |
| Varied pace | Plodding | | |
| Varied pitch | Monotonous | | |
| Emotional | Emotionless | | |
| Natural | Strained | | |
| Friendly | Unfriendly | | |
| Projected | Mumbled | | |
| Open mouth | Tight mouth | | |
| Crisp and clear | Lazy | | |
| Smooth | Jerky | | |
| Low and full pitch | High pitch | | |
| Comfortable pace | Too fast/slow | | |

# The dance

Body language is the strongest form of communication when in front of others. This is because it can create so many different perceptions and meanings without us even knowing it – unless we are more aware of it.

As a facilitator, all eyes will be on you. You need to know in advance the power you hold in creating the right environment just by the way you are standing and walking around the room.

An individual's perception of you will begin in a matter of seconds, based on your appearance, your voice and your body language. These perceptions will come from external factors such as their upbringing, experiences and general view of how they see the world.

First impressions are very important. We do have time, of course, to change that impression, as they do not know about our skills, subject-knowledge, etc., but a positive effect means that you can start with confidence.

### brilliant tip

Consider telling a story at the beginning of your workshop. It may make people more open-minded, engaged and focused on your words rather than on their pre-conceived first impressions.

## Face

- Smile naturally as you catch someone's eye – they will find it difficult not to respond in a similarly friendly fashion.
- 'Sweep' the group slowly with your eyes at frequent intervals.
- Use eye contact to read participants' body language – for example, for indications that they do not understand what you are saying.
- Resume eye contact with the group between glancing at your notes.
- Maintain eye contact with everyone. If you keep looking at the people on one side of the room the others will feel excluded.

## Posture

- Stand up in more formal situations.
- Stand to show (and feel) natural authority.
- Own your space.
- Sit down in more informal situations (but not behind a barrier like a table).

## Gestures

- Keep them natural – natural gestures move along a curve, not like a robot.

- Avoid repetition and irritating mannerisms.

- If you don't find gestures natural, then stillness is appropriate.

- Use gestures to emphasise key points and recapture the group's attention.

- Keep gestures to the top half of the body – they should add, not detract.

## The words

Think back to a time when you attended a workshop or event. The chances are that you can more easily remember how it made you feel, the people you met, the facilitator's style/approach or the venue, rather than recall what was said. You could probably pick out key messages or phrases, but when thousands of words are transmitted we actually place more emphasis on the experience.

This does not mean that what you say is not important, but how you say it can make the difference. Deliver the key messages at the beginning, middle and end. Repeating these messages over the session will give the audience the hook that they may need to remember a phrase or sentence that was crucial. Ask them to remember large chunks of the words used and they will find it difficult.

## brilliant example

Why is the music and dance memorable? I recall seeing a DVD of an 'inspirational' speaker and it captured members of the audience as they were leaving the venue. The interviewer firstly asked them how it was, and they said it was brilliant, the best thing ever, they felt so motivated and excited about facing the future and reaching their goals. He then asked them what the speaker had said. They looked at each other and laughed. They then responded by saying, 'I don't know!' Then they carried on by saying, 'But the way he said it and the passion he showed made me feel I could achieve anything.'

## Visual aids

*If a picture paints a thousand words...*

The above saying is so very true for a trainer/facilitator – or really for anyone who finds themselves in front of a group of people. Adding visuals to accentuate a point or enhance your argument can make all the difference.

There are a few areas to be aware of when using visuals. There are two particular traps that many fall into, among others.

The first is to be aware of talking to the slide or flipchart. You can get lost in the moment and focus on everything else but the audience. This is quite common. Slides should support you, not the other way round. The other point to remember is that your voice can be lost when talking in a different direction. You will need to increase the volume if your head is turned the other way.

## brilliant tip

Stand sideways to the audience if you are showing slides so you only need to rotate your head.

- Use visual aids to achieve something in your presentation that you can't do as effectively with words alone.

- A visual aid is a **means** of explaining ideas, not an end in itself.

- Go through the session and decide which information is difficult to get across just using words.

- Well-prepared visuals can add prestige and boost audience confidence in your authority.

- They must be appropriate and suitable both for the venue and for the size of the audience.

- Keep them simple and bold, uncluttered and clean.

- Keep them as **visual** as possible – use key words not prose; use illustrations or cartoons if appropriate.

- Ensure that they are shown long enough for understanding but not too long to be distracting.

- Use charts or graphs to show:
  - proportions;
  - contrasts;
  - developments;
  - sequences.

- Do not use them as a refuge – speak to the audience, not to the visual aid.

- Do not talk over the visual aid – you should allow your audience to absorb them, not insult your audience by reading the visual aid to them.

- Decide which is more important at any one time – the words or the images.

- Do not use too many words or display too many headings at once – the audience's natural curiosity will lead them to read on ahead of you.

## Using flipcharts

The preference for some facilitators to use flipcharts over PowerPoint slides is one of technology. There is far less risk using flipcharts because, as long as there is one present in the room, you simply need to ensure that you have markers that work, obviously in advance.

It is not that simple though, as room size, audience and topic need to be taken into account.

 **tip**

Practise writing on a flipchart. What can look acceptable to us close up may not to the person sitting far away. Therefore, check what size your writing needs to be for the person sitting furthest from the flipchart.

*Preparation*

- You may choose to arrive early to write up some pages of the flipchart in advance of the workshop. Alternatively, write them up a few days before.

- The flipchart should be used to support your talk – it should not be the centre of it.

- If you are reliant on the venue's materials, check the pens work! A variety of pen colours makes it more interesting for the audience.

- You may decide to use two flipcharts. For example, one flipchart might be used to write up a list of 'Dos' and the other flipchart to write up a list of 'Don'ts'.

- When setting up the room, check that everyone will be able to see the flipchart(s).

- Practise tearing off pages of flipchart paper without creating confetti!

*Presentation*

- Once again, do not talk to the flipchart as your audience will not be able to hear you.

- If you are writing up the participants' comments be sure to use their words. If the words they use are complex or difficult to understand ask them to paraphrase more simply. If they look to you for guidance suggest a version but ensure they are happy with it before you write it up.

- Most people feel comfortable standing beside the flipchart so that they can point to, or write up, words easily. Avoid hanging on to the flipchart.

- Use pre-prepared quotes or themed topics to put around the room. For example, on a leadership workshop, place leadership quotes around the room. This always creates conversations.

### brilliant tip

Write lightly in pencil in advance on the flipchart. When it comes to writing in marker pen you can just go over the pencil lines. This will help you to remember certain concepts/models, etc.

## Using notes

There is a perception among some people that when a facilitator looks at their notes they are demonstrating a sign of weakness, that they do not know their subject or they lack confidence. However, using notes or prompt cards is essential.

This is not an exam we are sitting, whereby we need to remember everything. The stakes are high. You have a group of people in front of you who are relying on you to deliver an effective learning experience for them. While they are with you, they are not doing their 'day job'. This is costing the organisation money so you need to demonstrate your worth.

There are a number of options:

- write a script out in full and read it;
- write the beginning and end to ensure a good start and finish;
- reduce the script to outline notes;
- reduce the script to skeleton notes;
- throw the script and the notes away!

The audience may like to see some notes – it shows you have done some preparation and promises that there will be an end to the presentation. They are also useful to instil discipline into the presentation and to protect you from rambling. Use them as a guide for:

- key points;
- when/how to use visual aids;
- timings.

They will help you to:

- retain your flow;
- stick to the logic of your presentation;
- make sure you do not leave anything out.

The type of notes you use will depend on your personal preference and the type of talk you are giving.

## Using a script

Writing a script may be useful as a starting point for writing notes. It may also give you confidence. You may want to read out a section word-for-word as it is imperative that you get it exactly right.

Beware of:

- reading verbatim;
- losing eye contact with your audience;

- losing your voice projection;
- losing your place;
- losing spontaneity and naturalness.

## Using prompt cards

These are very commonly used. They are a series of cards that act as your milestones throughout the session. They can be used to assist you through a day's workshop and to get you through your first few words. It can give you confidence to know that help is in your hand if you get stuck.

The writing on prompt cards can be anything you wish. It can range from session titles to paragraphs. In my research for this book, many colleagues remarked that having prompt cards to hold ensured that they didn't wave their hands around, and the holding of the cards gave them more confidence, knowing they were there to fall back on.

The prompt cards may:

- be less obtrusive;
- create a better flow;
- act as a clear guide.

Beware of:

- drying up and not recalling the messages behind the key point;
- getting them muddled up.

**brilliant** tip

Number your prompt cards and use a treasury tag to keep them in order. This will help you to return to the correct place if you happen to drop them.

## Using PowerPoint

There has been much said about the use of PowerPoint. There are compelling reasons both to use it and not to use it. An array of colourful, dynamic slides will in no way assure you of success. I have been witness to slides that contain over three different fonts, colours and pictures and my senses have been 'overloaded' and I have lost focus.

**brilliant tip**

Never have lots of words and paragraphs on the screen. Tell the audience what you need to and have a few bullet points to underline what you are saying. Use the 6 x 6 rule: no more than six lines per slide and no more than six words per line.

- Use diagrams/tables/pictures/flow charts wherever possible.
- Use colour to emphasise elements.
- Don't have too many words – have lots of background space on the slide.
- Use upper and lower case lettering.
- Keep layout consistent.
- Select images that will attract emotion – quality pictures over clip art.

Beware of:

- using too many 'word slides' (not really visual aids);
- creating boredom;
- equipment failure which may sabotage the presentation if you are relying totally on visual aids;
- spelling, grammar and punctuation.

*Pros and cons of PowerPoint*

| Pros | Cons |
| --- | --- |
| Can enhance clarity, interest and retention of focus. | A set structure can appear rigid. |
| Can illustrate complicated pictorial, statistical or conceptual material. | Dependence on technology. |
| Can give confidence if you fear public speaking. | Fancy fonts can be offputting and hard to read. |
| Can reinforce a message. | Sound of dynamic slide animations can be annoying. |

## The venue

Continuing with the music theme, the venue, the layout and the whole environment is your stage. The value of fully knowing your stage – where the flipcharts are placed, where the entrances are, how the seats are arranged – cannot be underestimated in building confidence. As a band, the sound-check lets you hear how you sound; adjusting guitar volume levels, knowing your side of the stage and adjusting your microphone levels so you can be heard. You feel more comfortable knowing you have been through a couple of songs and you have become familiar with your surroundings.

When you look at the venue for delivering workshops, you need to look at the general venue and then at the room layout and the options available to you.

So, get to know the venue before the workshop, if at all possible. If you are nearby or it is in your building then it makes your life easier. If you are travelling to the venue from a long distance, then ensure you arrive early and give yourself the time you need to manage your stage. For example, if you have planned to deliver your workshop standing up, ensure you have arranged sufficient space to move.

 **example**

Many years ago I was attending a workshop on effective presentation skills. The facilitator arrived just twenty minutes before the start and the majority of the participants were already in their places. As the room layout was quite cramped at the front, the first part of the day was ineffective as the facilitator spent their time demonstrating their presentation skills by squeezing through gaps between the tables and knocking into them. At lunchtime we agreed to move the tables to the back of the room so that the trainer had more space to demonstrate. Your environment plays a huge part in your success. Always arrive early to manage it.

If you are using PowerPoint or flipcharts, etc., check that each person in your audience will be able to see you and the visual aids.

The biggest fear when using technology is that it can fail you. Expect the unexpected. Check that the equipment is working and that you know how to operate it. Know where you can get spare bulbs, or know the person who you need to call and how you can contact them urgently.

Check levels of light and heat. If a room is too bright and too cold, it can seem threatening. Likewise, if a room is too dark and too warm, it can be an invitation to sleep!

Remove any unnecessary distractions – other people's charts, paper or equipment.

## Room options

A checklist

1   What sort of environment do you want to create? Any themes?

2   How do you want to lay out the room (circular, horseshoe, informal)?

3   What size room do you need? (Consider participants, exercises.)

4   How many rooms do you need (breakout rooms, coffee/tea)?

5   Where do you want people to sit/stand (speakers, managers)?

6   What could distract the participants (windows, noise, mobiles)?

7   Where do you want equipment to be placed (flipchart, projector)?

These various questions are explored below.

## What sort of environment do you want to create?

The environment, as we now know, is a very important consideration. You may be limited as there may be a small room allocated, or an organisation or department does not want anything moved. Think about the walls, placing pictures or quotes or questions on flipcharts around the room. Remember your stress balls/sweets/fruit, etc., and think long and hard about whether it is essential or desirable.

## How do you want to lay out the room?

The layout of the room, if you have the freedom to do so, is another important decision you will have to make. There are many different options available to you here. Some of the most popular layouts are shown on the following pages (see Figures 7.1–7.4), together with the advantages and disadvantages of each.

*Horseshoe*

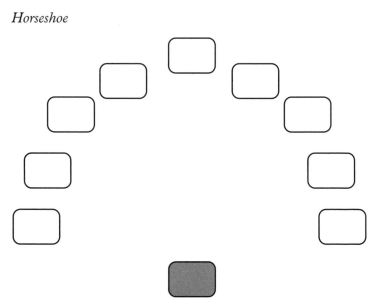

**Figure 7.1** The horseshoe layout

| Advantages | Disadvantages |
|---|---|
| Removes obstacles. | Can seem like a 'self-help' group. |
| Everyone deemed equal. | Nothing to lean on if you need to write (you can offer clipboards, etc.). |
| Relaxed atmosphere. | Can be too relaxed. |

*Boardroom*

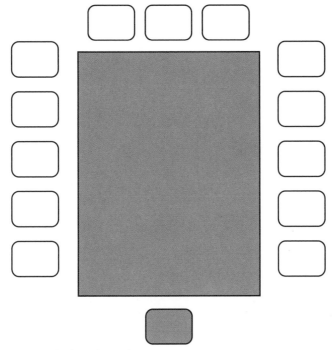

**Figure 7.2** The boardroom layout

| Advantages | Disadvantages |
| --- | --- |
| Participants comfortable as familiar with layout. | May see the back of someone's head when viewing slides, etc. |
| Everyone deemed equal. | Can be quite rigid – lack of interaction with others in the room. |
| Everyone has their 'area'. | Can seem like a formal meeting. |

*Cabaret style*

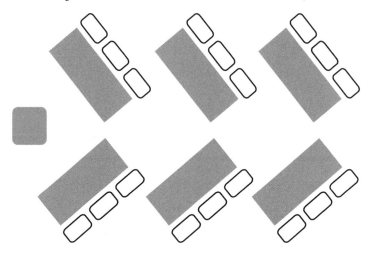

**Figure 7.3** The cabaret-style layout

| Advantages | Disadvantages |
| --- | --- |
| Participants sitting in smaller groups of two or three. | People sit with their colleagues – lack of mixing with others. |
| Everyone deemed equal. | Room needs to be big enough to accommodate as takes up more room per participant. |
| Different room layout to the norm can create a positive beginning. | Can seem like an assault course to get through for a facilitator. |

## Circular

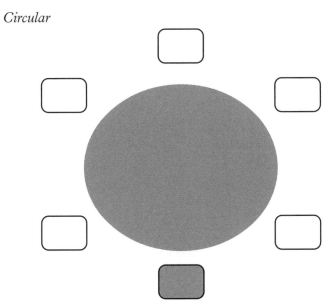

**Figure 7.4** The circular layout

| Advantages | Disadvantages |
| --- | --- |
| Easy to address everyone. | Can be too informal. |
| Everyone deemed equal. | Can feel 'claustrophobic' if a long session. |
| Table not a barrier. | No front focal point of room for slides/ flipchart, etc. |

## What size room do you need?

The bigger the room, the more flexibility you will have when arranging exercises and considering the seating places of the participants. When designing the workshop, the size of the room should be a consideration as it will determine how interactive your sessions could be.

 **example**

A colleague once told me that she had designed a team 'away-day' with lots of interaction and moving around in smaller groups to get to know each other. She was led to believe the room would be big enough. However, on arrival she found a small room with chairs equally close to both the table and the walls. She realised that she would have to quickly rethink her strategy. It taught her to get specific room sizes in future, as everyone's perception is different.

You can always work with a room that is too big for the participants. You can separate it off into sections; have different parts of the room for thinking/exercises/discussion, etc. If, on arrival, you find that the room is too small, for whatever reason, you will have to adjust accordingly. On no account spend the rest of the workshop complaining about the size of the room. Mention it once, then move on and deliver with the tools you have available.

## How many rooms do you need?

You may want to have some breakout rooms so that smaller group work can be done in a little more privacy. Do you need a separate room for tea/coffee/networking? By having breakout rooms, it can give you some time on your own to set the room up for the next session as well as have some valuable thinking time.

## Where do you want people to sit/stand?

If you have invited a speaker, then ensure that you have a seat for them and that they can address the group from a central position. If you know the audience you may want to put out placecards so that you can manage group dynamics by keeping apart those who are likely to be a challenge, for example.

## What could distract the participants?

Try to reduce the distractions as much as possible. Obviously some will be out of your control. If the window leads straight onto the street then every person who walks by will be a potential distraction as people look up. Use blinds if they are there. Also, don't stand in front of a ground floor window if it is a throughway, as people will be constantly looking over your shoulder. Manage room temperature against potential increase in noise if the windows are open. If in doubt, ask the group for their preference.

## Where do you want equipment to be placed?

Many rooms have their projector set from the ceiling so flexibility is limited when presenting slides. If you are using your laptop and a portable projector, get there early and ensure that the picture is clear and it is enlarged so that everyone can see it. If you are just using flipcharts, you have more choices available. I sometimes place one at each end of the room when standing at the 'front' is not imperative.

# Your style vs subject matter

It is important to match *what* you are saying to *how* you are saying it. If you are running a workshop on public speaking and your voice is very quiet and your body language wooden, then in all probability it will not be received very well.

Think about the following subjects and *how* you would deliver them:

- An introduction to a sickness absence process.
- Presentation skills.
- Stress management.

The table below shows some mixed individual styles that you may want to match up to the relevant workshop. There are, of course, many others.

| Tell style | This is how it's going to be! | Sell style |
|---|---|---|
| Demonstrate technical ability | Enabling style | Loud and strong |
| Soft and understanding | Empathetic | Directive |

It has been said that to deliver an effective leadership workshop, you need to be able to demonstrate strong leadership in taking the participants forward. Remember that leadership by its nature can be directive or reflective, rigid or flexible, pessimistic or optimistic. You will need to be flexible in your approach, just like a leader is at different times. Your style should be situational.

**brilliant** recap

- Have confidence – you are there for a reason.
- Write three key messages you want the audience to take away and build from there.
- Balance your words, music and dance.
- Choose visual aids that are relevant.
- Adapt the room layout for the needs of the group.

# PART 2

Onstage – stand and deliver (during the workshop)

Thils section is all about the actual delivery. We have planned, practised and prepared and now the curtain is about to go up. We need to be ready. We have thought of all possible scenarios (generally worst-case ones!) but still this is the part that holds the most fear for many people.

## It's my turn!

This is it. It is you in the spotlight. Chapter 8 will focus on getting you off to a great start. You need to be in control (on the outside at least) and enable your audience to see that you know what you are doing and that it is going to be fun/exciting/exhilarating/dynamic (delete as necessary). Your body language and tone will be your best friends and you will be aware of how to use them to your advantage.

## Are the participants engaged?

Chapter 9 will look at managing group dynamics. Two or more people make a group. You need to have a range of strategies available to identify certain 'types' and then have the solutions to deal with the situation and the person.

## How do I know if it's 'hitting the spot'?

Chapter 10 will give you checkpoints that you can use to see if everything is going as you had hoped it would. Times of the day have an impact on energy levels and we must adjust our styles accordingly. Ongoing self-assessment and evaluation can help you to gauge progress and the 'people temperature'.

# CHAPTER 8

'And now I'll pass you over to…'

This is it. The lights go down and the curtain goes up. All the planning, questions and feelings you have are now taken over by the sound of your heart beating faster – like a bass drum that we think everyone in the room can hear. We are self-conscious but want to seem calm and in control. In this chapter we look at getting you off to a great start. Body language and your voice are key – and you have to manage them accordingly. We will focus on controlling any nerves, breathing techniques, memorising words, as well as identifying the benefits of standing vs sitting and moving vs remaining still.

 We never get a second chance to make a first impression.

Irma Wyman

Wow. All that planning and now here we are, just about to say our first words in front of our audience. Our logic is replaced by our emotions and suddenly we can feel vulnerable and question why we agreed to do this in the first place.

Rather than ramble on, which is easy, it can be beneficial to take the participants through a simple model about the context in which they will be working for the session/day (see Figure 8.1 overleaf). It will get you to become comfortable with your voice being heard and it is a model that is common sense and non-threatening. In 10 years of using it, all I have heard is that it is common sense.

## The challenge vs support model

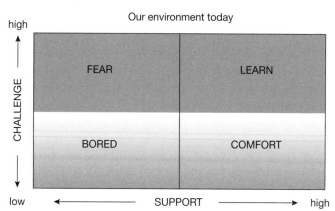

**Figure 8.1** The challenge vs support model

As you can see from Figure 8.1, we have challenge on the vertical axis and support on the horizontal axis. Low and high are at each end of the line. If showing it on PowerPoint, it is very effective if you can use custom animation so that only one word appears at a time.

Show the four-quadrant model on a screen or flipchart without any of the words in the quadrants showing.

Then show/write the first word, BORED. Explain that if there is low challenge and low support, then people will get BORED easily. Speaking as their facilitator, I do not want this.

Then show/write the word, COMFORT. Explain that if there is low challenge with high support from others, then people will be in their comfort zone and it will be a lost opportunity. We do not want this.

Then show/write the word, FEAR. Explain that if there is a lot of challenging by others but no support mechanism, then people will be reluctant to talk out through fear. Again, we do not want this.

Where we want to be is in the top right quadrant, which will be conducive to get people to LEARN. If we encourage high challenge with high support, then people will be away from fear, boredom and their comfort zone and will truly open themselves up to learn. This is where we want to be!

Put it to the group and get agreement. This exercise will show them that you are in control and your voice has been assertive in defining the environment you want to create.

## Controlling nerves

 If you want to test your memory, try to recall what you were worrying about one year ago today.

E. Joseph Cossman

- Most people fear presenting in case they should look foolish or dry up. Yet, you can prevent both of these from happening by being prepared. Remember that 90 per cent of your presentation should be preparation.

- You will probably find you are most nervous in the few minutes before your talk, the first minute of your talk and when handling questions. In the minutes before try breathing deeply, and when you are at the front stand tall; both of these will increase the oxygen taken in to your lungs. Also, drive yourself to want to speak.

- Remember that audiences are generally open and accepting. They are also human. Look at the people in your audience as friends, so talk to your audience informally before your presentation. That is also helpful for personalising your talk to your audience.

- To reduce your nervousness in the opening minutes it is best to have learnt your opening sentences. Speak them with courage and confidence, and think of yourself as there to succeed.

- It is difficult to do much to reduce the nerves of answering questions other than to have anticipated questions beforehand. Try not to let your nervousness show when answering questions. Consciously relax your facial expressions.

- Ensure your body language does not show your nerves. Relax your facial expressions and your arms. Do not clench your hands.

- Be calm so that you can deal with the unexpected. Do not worry about trivialities.

- Do not panic, or take yourself too seriously, if you make a mistake during your talk. We are all human.

- It helps if you dress smartly.

- Do not try to be someone you are not. Know yourself, like yourself and be yourself.

**brilliant** tip

When you are listening to individuals introduce themselves, repeat their name back to them. This will help you to memorise their names as well as get clarity on pronunciation.

Successful theatrical productions are rehearsed many times before the first performance and giving a presentation is no different. As well as using rehearsals to perfect your presentation, you will become familiar with the content and structure, which will inspire confidence and in turn reduce nerves. Practise on your own, practise in front of a mirror, and if you can find someone to watch your presentation then you'll be far more comfortable when you come to the real thing.

## But what about those last-minute nerves?

A thorough rehearsal will significantly reduce your tension; however, it is inevitable that you will still feel some nervousness immediately before your presentation. Here are a few techniques that will help you to deal with those last-minute butterflies.

### Internal warm-up

Visualise yourself giving a successful presentation. See the audience listening and responding positively to your words. Hear their laughter as you make a joke. Listen to the conversations they are having. Hear their applause. Run through your opening lines in your head.

### Physical warm-up

Professional actors will often run through a basic physical warm-up in the wings before they make their opening entrance. Physical activity is a great way to reduce tension.

- Stretch out your body. Short five- to ten-second stretches will help to reduce tension.
- Clench your fists and toes tightly and then release. Repeat this action five or six times.
- If possible find a space to do four or five press-ups. This will reduce upper body tension.

### Breathing

- Take a deep breath down into your stomach, hold for 10 seconds and then release the breath counting aloud until all of the air has dispersed. Repeat two or three times.
- Take a deep breath through your nose then breathe out slowly through your mouth. Repeat this action two or three times.
- Taking control of your breathing will improve your voice quality.

*Beware of what you consume*

- You should avoid any products which will dehydrate you, resulting in your throat and mouth feeling dry. It is also wise not to eat a very large meal ahead of a presentation. This can lead to you feeling sluggish and lethargic.

- You will have a natural flow of nervous energy and therefore will not require caffeine and sugar to give you an additional boost.

- Drink a sensible amount of water which is served at room temperature. Too much water will increase your bladder activity and add to the stress of the situation!

## Memorising words

When working with a set text, the most effective method of committing the words to memory is through repetition. If you have the time, practise the words at the same time each day for the same length of time, preferably for between 30 minutes and 1 hour. This will lock your presentation into your brain and give you the confidence to perform without notes.

## Enthusiasm

- Be enthusiastic, as it is contagious.
- Use gestures (usually hands) to emphasise your enthusiasm.
- Project your personality. This warms the audience to you.
- Be careful with humour. Some people get offended very easily.
- Sell your message with confidence and conviction. Do not oversell by inundating them with facts.
- Be sure to be enthusiastic even if you have given the talk many times already. You need the audience to feel that they are on the receiving end of your enthusiasm and that your delivery style is fresh and vibrant.

## Distractions

- Expect distractions from latecomers, splinter groups, ill-timed questions or loud noises from outside the room. If you think the loud noise will be short-lived, for example, an ambulance siren, you should pause rather than shout over it.

- You can avoid ill-timed questions if you say at the beginning that either you would like to answer questions at the end, or you will invite questions at intervals. Alternatively, defer answering the ill-timed question until you have finished making a point.

- You may not be successful in keeping the attention of all your audience. If anyone is disruptive you could leave it to those around them to quieten them, or you could use firm, steady eye contact with them. If they are not disruptive, ignore them. In the break talk to them to identify the hidden agenda that they are disappointed you are not addressing and aim to change their attitude and understanding.

## Standing up or sitting down?

Let's get this straight. When first getting the courage to speak in front of other people, it is a lot easier to sit down than to stand up in front of them. There is nothing wrong with that, if it builds up your confidence.

When delivering your workshop, you will mostly be required to be on your feet. Being flexible, of course, you will spend some time sitting down as well and there are times when this is appropriate.

## When you may want to stand up

● To make a positive impact when saying your first words.

● To control the group dynamics.

● So that you can scan the room and participants.

● So you can build rapport through eye contact.

● So that you can make your body language visible.

● To make a point.

● To manage a cross-group discussion.

● To make an announcement.

Be aware of:

● being too animated;

● moving up and down the room like a yo-yo;

● any mannerisms you may have – such as flicking a pen top or pulling at Blu-Tack or paper clips.

## When you may want to sit down

● To be equal with the group.

● To be part of a discussion.

● To interact without authority.

● To have a conversation.

● To engage at their level.

● To show an interest.

● To listen attentively.

● To turn attention away from you.

Be aware of:

● wanting to be one of 'them';

● hiding behind the table and therefore hiding a large percentage of your body language;

● losing your authority and control of the environment.

## brilliant recap

- Memorise your first words.
- Use tools to help you, such as the challenge vs support model.
- Control any nerves by undertaking breathing exercises.
- Be enthusiastic – it can be contagious.
- Be yourself.
- Lastly, remember that *you* create the environment, not the group.

# CHAPTER 9

# Managing group dynamics

roup dynamics are present when there are three or more people. This chapter focuses on what they are and how to identify them. By knowing the different types, you will be able to create exercises that fit into their particular preferences. We will look at managing challenging people and what we must do to keep the rest of the group safe by deciding on ground rules or boundaries. We will also look at 10 typical challenging behaviours and what you can do to handle them.

It is your role as trainer or facilitator to make sure that the group feel safe in their environment. Generally, you will be providing the 'what' of the workshop and the group will provide the 'how'. You may have the outline, aims and objectives, but it can be useful to hear from individuals how they would like to achieve them.

Although as a facilitator we may feel nervous or anxious before we start, this can also be said of many of the participants. One of the key ways a facilitator can manage this is by helping the audience establish ground rules.

## Ground rules

Ground rules are the rules of conduct or behavioural guidelines that members of the group agree on before you proceed with the workshop. They are based on an assumption of equality and fairness. The idea is that no individual is permitted to dominate a discussion or hold special privilege.

There are generally three kinds of ground rule:

1 The first kind defines the **behaviour** of participants – for example, 'Individuals will treat each other with respect.'

2 The second kind applies to **procedures** to be used by the group – such as, 'All decisions will be made by consensus.'

3 The last kind of ground rule may also define the **boundaries** of discussions on certain issues – for example, 'Discussion today will focus solely on the issue of using the new appraisal system, and will not go into discussions of the old systems.'

Try practising agreeing ground rules with a group. Some facilitators who have run many effective workshops know exactly the types of behaviour, procedures and boundaries needed for the day to be effective so will therefore inform the group of what is required.

The majority of the time, if you work on ground rules with a group, you will look to 'agree' them, rather than 'set' them. If the rules come from the group they are more likely to stick to them than if they come from someone else.

**brilliant** timesaver

When going round the room and getting introductions from individuals, at the same time ask them for one ground rule each. You can write this rule on a flipchart and place it for all to see for the session. This is quicker than having a separate 'ground rules' exercise, which can eat into your time.

## When to use

● When working with unfamiliar groups, departments or organisations.

- If the audience is constantly being distracted.
- To allay any fears or anxiety the group (or facilitator) may have.
- If the audience is non-participative (quiet).
- When there are sensitivities around group members or the subject.

## When not to use

- With a familiar group – familiar either to you or themselves.
- Just for the sake of it at every workshop.
- When you want the group to be creative.

### brilliant tip

You can refer to ground rules at any time. It doesn't just have to be at the beginning. If you feel that you are losing the audience because of a dominant member or too much talking, then run a ground rules session to get things back on track.

# Asking and answering questions

In Chapter 1 we looked at questions to ask a client to identify what is required. One skill that a facilitator needs to focus on is the asking and answering of questions in front of a group. A common fear is being able to answer questions. However, you also need to be able to ask your audience questions.

## Asking for questions

How a question is phrased will depend on the reason for asking it. It might be for one of the following reasons:

- To check understanding.
- To start a discussion.

- To keep participants alert.
- To encourage participation.
- To pool ideas.
- To get feedback.

Decide when you will ask for questions: during and/or after your sessions. The advantage of allowing questions during your talk is that your audience are more likely to listen to, and think about, what you are saying.

If you choose to leave questions until the end, you can stimulate a more lively questions and answers session by asking rhetorical questions during your talk.

Select an appropriate tone of voice that will encourage your audience to ask questions. Decide how you want the person who is asking the question to identify themselves. Will you ask them to stand up or would you like them to put up their hand? You can achieve the latter simply by raising your hand when you ask if anyone has a question.

## Listening to questions

Identify why a question is being asked. If it is to clarify something you said earlier, look for signs from the client or other people not having understood either. This will indicate how full an answer to give. It will also give you general feedback on your audience.

Somebody may ask for more information simply to satisfy their own curiosity. Try to gauge other people's interest too, as this will help you decide how much information to give.

Others may ask a question to prove something either to you or to others in the audience. If you can identify what they are trying to prove, it can help you answer appropriately.

Be sure that you understand the question. Avoid showing surprise at the question asked, even if it appears that the person has not been listening to what you have been saying. Also, do not laugh at the question, even if the rest of the audience do. Give your whole attention when you are asked a question.

- Listen to the question, observe what is **not** said and sense the emotions of the person asking the question.

- When listening to the question, remember to relax your facial expressions. Appear calm.

- You should repeat the question you have been asked so that the rest of the audience can hear it. It also gives you more time to think about it.

- Always listen to a question patiently, however lengthy it is. When asked a long question, try to identify the key question within it.

Identify the type of person who is asking the question as this will affect how you handle it. Here are a few suggestions:

1 **The complainer**: Listen to the complaint. Explain that you are unable to do anything during the session. Move on.

2 **The obstinate person**: Give an answer, then say that time is short but that you would be happy to discuss the point with them later.

3 **The argumentative person**: Remain calm. Try to agree with something that they say. Win them round at the break.

4 **The opinionated person**: Restate your views. There is no reason why you should have to concur or disagree, but recognise these are differences of opinion, not fact.

5 **The informer**: Acknowledge their points and say that you will keep them in mind.

## Answering questions

It is OK not to know every answer. Facilitators put themselves under so much pressure by being concerned that they will look foolish or silly if they cannot answer a question on the subject they are discussing. There will always be someone who knows something (or thinks they know something) more than us. It is a learning experience for the facilitator as well as the audience.

Manage expectations by informing the audience that you are not the expert. Explain that you do not know all the answers. Your role is to help them find their own answers. Obviously, if you are talking about a process that you have designed that you want others to follow, then you will be deemed to be the 'expert' and you should know what you are talking about. Be very clear about your role and the expectations of the audience.

- Take time to plan your answer.

- Keep your answer simple and concise. If a longer answer is required, defer giving it until the break.

- Make your answer relevant to the audience. It may help to add an example to illustrate your point.

- If you are pressed for an opinion, limit your risk by supporting it with facts.

- Answer in a sensitive and helpful manner. Others will be watching you to see how you handle questions before asking one themselves.

- Avoid being defensive, controversial for the sake of it or aggressive.

- If you do not know the answer to a question, say so, confidently. That way you will preserve your credibility.

- Generally, you should check whether you have answered the question. However, if you have a complainer or an argumentative person asking the question, you may do better to give a brief answer, then swiftly move on.

- Having answered a question during your talk, you may have lost your train of thought. It is even more likely that the audience has, so summarise up until the question was asked.

## Creating the right ambience

Group dynamics is the study of groups, and also a general term for group processes. Because they interact and influence each other, groups develop a number of dynamic processes that separate them from a random collection of individuals. These processes include norms, roles, relations, development, need to belong, social influence and effects on behaviour.

The basic skills of a facilitator are about following good practices: timekeeping; helping people through agreed aims and objectives; and providing necessary breaks. The higher-order skills involve watching the group and its individuals in light of group process and dynamics.

A facilitator needs to be able to assist a group in accomplishing its objective by diagnosing how well the group is functioning as a problem-solving or decision-making entity and intervening, where necessary, to alter the group's operating behaviour.

For example, a facilitator should look out for:

- patterns of communication and coordination;
- patterns of influence;
- patterns of dominance (e.g. who leads, who defers);
- the balance of task focus vs social focus;
- the level of group effectiveness;
- how conflict is handled.

The role of the facilitator is to continually draw the group's attention to the group process and to suggest structures and practices to support and enhance the group skills.

## Tuckman's model of team development

The best-known model of groups or teams is that of Bruce Tuckman (1965), who identified four stages of team development (see Figure 9.1). Tuckman described the four distinct stages that a team or group can go through as it comes together and starts to operate and perform. This process can be subconscious, although an understanding of the stages can help groups reach effectiveness more quickly and less painfully.

There is no time allocated to any one stage. A group or team can move very quickly through the stages or can remain in one for a long time. A group may also start at any of the stages. It depends on the members and the situation attached.

As a facilitator you will face groups of individuals operating at different stages of this model, and understanding how to recognise the stage of the group will enable you to 'flex' your style accordingly. It is unlikely, but not impossible, that a group of people will go through each stage of the model within one brief session, but your task as a facilitator is to enable them to 'perform' effectively during the workshop.

1 Forming: pretending to get on or get along with others.

2 Storming: letting down the politeness barrier and trying to get down to the issues even if tempers flare up.

3 Norming: getting used to each other and developing trust and productivity.

4 Performing: working in a group to a common goal on a highly efficient and cooperative basis.

It should be noted that this model refers to the overall pattern of the group, but of course individuals within a group work in different ways. If distrust persists, a group may never even get to the norming stage.

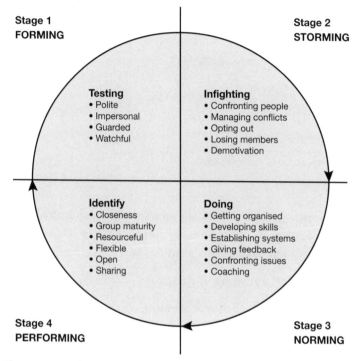

**Figure 9.1** Tuckman's model of team development
*Source*: Based on data from Tuckman (1965)

## The four stages

### Forming stage

When a group or a team get together there are certain feelings that can be associated with each individual. These can be both positive and negative, or indifferent. These feelings must be acknowledged if you want the group to perform at its best. This is also known as the 'identity' stage, whereby the group needs to know why it exists and what it is there to do.

Try to remember when you were part of a new project team or attended a training day with people you didn't know. How did you feel? Use some of the prompts from the table overleaf and/or add your own. Then try to remember why you had those feelings.

| My feelings/attitude | What did this tell me about myself? |
|---|---|
| Nervous | |
| Quieter than normal | |
| Guarded | |
| Wary | |
| Defensive | |
| Overcompensated to make an impact | |
| Excited | |

In the forming stage, individuals seek to establish personal identities and make an impression on other members of the group. They search out each other's attitudes and background.

People may feel a mixture of things:

● Excitement, anticipation, optimism, recognition.

● Discomfort at not knowing others or what is expected.

● Scepticism – 'Will they really listen to me?'

● Concern – 'What are the consequences of failing?'

When a group comes together for the first time, what skills and techniques should a facilitator demonstrate to enable the group to move to the next stage?

Get a pen and paper and make some notes on what a facilitator needs to demonstrate – the skills and approach required – to meet the group's needs at this time. Remember how you felt in the above exercise and think about what you would have wanted from a facilitator at this time.

### Storming stage

The storming stage is something that is perfectly normal in the evolution of a group. This is the 'infighting' stage. As a facilitator, there may be some confronting and jostling for position. Expect it. Once you expect it, you can plan for it. This is the toughest part to get through. Some groups stay forever at this stage. It can be draining and reduce morale, or it can be a place where people thrive on their competitive nature.

Can you remember being part of a team or attending a training day when there was uneasiness, infighting (overtly or covertly), people wanting to opt out, open confrontation, etc.? How did you feel? Use some of the prompts below and/or add your own. Then try to remember why you had those feelings.

| My feelings/attitude | What did this tell me about myself? |
|---|---|
| Frustrated | |
| Watching cliques form | |
| Wanted to opt out | |
| Challenged openly | |
| Became more introverted | |
| Where am I in the pecking order? | |
| See who is on my side | |

In the storming stage, personal goals are revealed which can lead to hostility and conflict. It is a bargaining phase in which individuals attempt to sort out group and individual objectives.

Members feel the full range of emotions:

- Anger and frustration at not understanding or being excluded.
- Fear and concern at workloads or dominant people in the group.
- Joy at success and getting things right.
- Disagreements with the way things are going.
- Misunderstandings of what people thought was happening.

When a group is in the storming stage, what skills and techniques should a facilitator demonstrate to enable the group to move to the next stage?

Get a pen and paper and make some notes on what a facilitator needs to demonstrate – the skills and approach required – to meet the group's needs at this time. Remember how you felt in the above exercise and think about what you would have wanted from a facilitator at this time.

*Norming stage*

When a team is in the norming stage, it is generally called 'the calm after the storm'. Individuals are comfortable as to why they are part of the group and are OK with the levels of hierarchy within it. Systems and procedures are being established, people can give feedback more freely and issues are dealt with in a more positive way. The group is more 'mature' and individuals are open to learning new skills.

Can you remember being part of a team or attending a training day when people accepted their roles within the group, and there was a willingness to adapt and a sharing of common values, etc.? How did you feel? Use some of the prompts below and/or add your own. Then try to remember why you had those feelings.

| My feelings/attitude | What did this tell me about myself? |
| --- | --- |
| I could positively challenge others | |
| I could openly communicate my ideas | |
| I could give and receive feedback | |
| I was actively listening | |
| There was a sense of cohesion | |
| I was actively involved in achieving group goals | |
| I felt a sense of team spirit in the group | |

The team develops ways of working in order to achieve objectives. Roles are allocated and norms of behaviour, i.e. working rules, are established. The effect is to create a framework within which team members can relate to each other.

Members feel a mixture of things:

- Task confidence is growing well.
- There is a stronger sense of team spirit.
- Confidence building is the real need.
- The group starts to feel it can achieve things without direction.
- The team looks to take on new challenges.

When a group is in the norming stage, what skills and techniques should a facilitator demonstrate to enable the group to continue to develop and move to the next stage?

Get a pen and paper and make some notes on what a facilitator needs to demonstrate – the skills and approach required – to meet the group's needs at this time. Remember how you felt in the above exercise and think about what you would have wanted from a facilitator at this time.

*Performing stage*
Wow! This is where it feels good. The group has matured and is now a fully functioning unit. Everyone is settled in their roles and looking to add value. Individuals are looking to the future rather than the past. It feels good to be a part of this group. There is support and trust and a feeling of being valued for your contribution.

Try to remember when you have been part of a team or attended a training day when people felt that the group gelled from the start. There was a high level of positive challenge going on, together with a high level of support. You felt valued and wanted to succeed with the others around the room. Use some of the prompts below and/or add your own. Then try to remember why you had those feelings.

| My feelings/attitude | What did this tell me about myself? |
|---|---|
| I was supportive to others | |
| Others were supportive to me | |
| Morale was high | |
| Responsibility was shared | |
| I was flexible to adapt to situations | |
| There was insight and innovation | |
| Celebration of what we achieved | |

The group, having resolved earlier problems and differences, can now get on with the task in hand, and work effectively and efficiently together.

Members have a number of needs:

● They require confirmation that the team is performing capably.
● They need to feel independent and that they are sharing leadership.
● They need to feel recognised for the roles they play.

Get a pen and paper and make some notes on what a facilitator needs to demonstrate – the skills and approach required – to meet the group's needs at this time. Remember how you felt in the above exercise and think about what you would have wanted from a facilitator.

## Facilitator styles for the four stages

Please remember that no two groups are the same. A group can come together and can go through the forming stage very quickly. It is situational so do not be rigid with your style and approach.

## Forming

When a group is together for the first time, there will be a range of individuals with different feelings. Depending on the subject matter and your objectives for the group, at these early stages of the group getting together, your style should be a *directive* one. They will need to know why they are there, what are the key timings and learning objectives and what you want from them.

## Storming

This stage is more prevalent in teams working in organisations for longer periods, but it can happen at workshops. Watch out for put-downs, animosity, rudeness and the quiet types. If this is occurring, the style we should use is either to be more *directive* (as above) to assert your authority if needed, or slightly more *supportive*, where you listen and empathise, and understand each person's viewpoint to get the group back on track.

## Norming

The group is working well. You do not need to be authoritative. You can step back in your style and just support the group in areas that they need. You can use a *coaching* style to ask lots of questions to enable them to come up with their own solutions.

## Performing

When your participants are at this stage of Tuckman's lifecyle, you can truly *facilitate* and nudge the group along to get them where they want to be. It is a pleasure working at this stage and the group can really benefit from the 'enabler' role that you take up.

In football terms, it is said that the referee has had a good game when he is hardly mentioned or noticed, and the game has flowed. There is some correlation to the role of a facilitator – the group (the teams) want to succeed and may not even notice that you have been there! You will have kept things going, having quiet words with individuals when required and monitoring group dynamics and potential barriers ahead.

## Group styles

The two theories below are particularly useful when looking at effectively understanding and managing group dynamics. Although there are many others, these are my personal favourites and have been helpful in developing my skills as a facilitator.

### Adaption-Innovation Inventory

Kirton (1989) identified two different styles in creativity, problem-solving and decision-making. The Adaption-Innovation Inventory and its associated psychometric instrument (KAI) can provide you with insight into how people solve problems and interact while decision-making. The Adaption-Innovation Inventory is founded on the assumption that all people solve problems and are creative. See **www.kaicentre.com** for more information.

- **Adaptors**: those who prefer to improve on existing practice.
- **Innovators**: those who prefer to reframe problems in a way that often confronts accepted practice.

It is important to recognise that both adaptors and innovators are required in a balanced group. Both styles are able to produce creative approaches and solutions, but the important thing to note is that they start from different places.

They both support the innovation process but just have different approaches, and it is recognition of the differences by the facilitator and the group that will help maximise the strengths of both styles. The facilitator will need to balance the two.

*Adaptors*

- Demonstrate precision and a methodical approach, erring on the side of prudence.
- Seek solutions using tried and well-understood methods.
- Rarely challenge the rules.
- Produce few new ideas.

*Innovators*

- Appear to be undisciplined and approach the problem from unusual angles.
- Query the basic assumptions around the problem – may appear abrasive and upsetting to others.
- Challenge the rules.
- Produce many ideas – some of which may appear risky.

As a facilitator, your role is to 'balance' these two styles if you have a mixed group of both adaptors and innovators, as well as to potentially counter-balance the styles when a group consists of mostly adaptors or mostly innovators.

## ↗ brilliant exercise

Identify three techniques to recognise and respond to these different styles when facilitating group dynamics.

| Adaptors | Innovators |
| --- | --- |
| | |

Ideas:

● Split the group when problem-solving or brainstorming.

● Divide the session up into two parts, for example:

– continuous improvements to existing processes;

– brand new ideas and solutions.

● Agree up front with the group whether the meeting is about new ideas or extensions to existing practices.

## The Bolton and Bolton model

The second theory is from Bolton and Bolton (1984). They identified four different social styles:

● **The Driver**: Task-oriented who attempt to influence by asserting themselves in a controlled way.

● **The Expressive**: They are assertive like the driver but use their feelings to help assert themselves.

● **The Amiable**: Openly show their feelings but less aggressive and assertive.

● **The Analytical**: Control their feelings and emotions and tend to be task-oriented, gathering and testing factual information.

Your group(s) can be broken down into the following 'styles' and your delivery and content adapted accordingly.

*The Driver – control specialist*

| | |
|---|---|
| Pushy | Determined |
| Severe | Requiring |
| Tough | Thorough |
| Dominating | Decisive |
| Harsh | Efficient |
| Results-oriented | |

## *The Expressive – social specialist*

| | |
|---|---|
| Manipulative | Personable |
| Excitable | Stimulating |
| Undisciplined | Enthusiastic |
| Reacting | Dramatic |
| Promotional | Gregarious |
| Applause-oriented | |

## *The Amiable – support specialist*

| | |
|---|---|
| Conforming | Supportive |
| Retiring | Respectful |
| Ingratiating | Willing |
| Dependent | Dependable |
| Emotional | Agreeable |
| Attention-oriented | |

## *The Analytical – technical specialist*

| | |
|---|---|
| Critical | Industrious |
| Indecisive | Persistent |
| Stuffy | Serious |
| Exacting | Vigilant |
| Moralistic | Orderly |
| Actively-oriented | |

As a facilitator, just be aware that there are many different dynamics that need to be taken into consideration. Know that each group will be different. This may be obvious with the demonstration of extroverted behaviours, or it may be subtle and you will need to use your senses to pick up on all possibilities.

*How to facilitate the different social behavioural styles*

## Driver:

- Be clear, specific, brief and to the point.
- Stick to business.
- Come prepared with all requirements, objectives and any support material in a well-organised 'package'.
- Present the facts logically; plan your presentation efficiently.
- Ask specific (preferably 'What?') questions.
- Provide alternatives and choices for making their decisions.
- Provide facts and figures about the probability of success or effectiveness of options.
- Motivate and persuade by referring to objectives and results.
- Support and maintain.
- After talking business, depart gracefully.

## Expressive:

- Plan interaction that supports their dreams and intuitions.
- Ensure enough time to be stimulating, fun loving, fast moving, entertaining.
- Leave time for relating, socialising.
- Talk about people and their goals and opinions that they find stimulating.
- Ask for their opinions/ideas regarding people.
- Provide ideas for implementing action.
- Provide testimonials from people that they see as important, prominent.
- Offer special, immediate and extra incentives for their willingness to take risks.

**Amiable:**

- Start (briefly) with a personal comment. Break the ice. Use time to be agreeable.

- Show sincere interest in them as people; find areas of common involvement. Be candid and open.

- Patiently draw out personal goals and work with them to help achieve these goals; listen/be responsive.

- Present your case softly, non-threateningly.

- Ask 'How?' questions to draw their opinions.

- Move casually/informally.

- Define clearly individual contribution.

**Analytical:**

- Approach them in a straightforward way; stick to business.

- Support their principles/thoughtful approach. Build your credibility by listing pros and cons to any suggestion they make.

- Make an organised contribution to their efforts, present specifics and do what you say you can do.

- Take your time but be persistent.

- Draw up a scheduled approach to implementing action with a step-by-step timetable; assure them there won't be any surprises.

- Give the time to verify the reliability of their actions; be accurate, realistic.

## Dealing with challenging individuals

As well as looking at the group dynamics, we need to have the ability to identify individuals who may be the cause of these dynamics or who have personal or political agendas, or just the individual who is very quiet and needs a different approach.

Everyone attending a workshop has either paid for it themselves, or their organisation has paid. Payment does not just have to be financial – there is also a big payment in time. Time out from the 'day job' means that their mind may be elsewhere. The role of the facilitator is to ensure that they can make the link between being participative and positive and how it will benefit them in the workplace.

There are many different types of individual you may face. They can range from aggressive to passive. The table below looks at 10 types of person, their typical behaviours and what should be done to manage them effectively.

| AGGRESSIVE | |
|---|---|
| Description | Pushy and aggressive. Tries to dominate. |
| Typical behaviours | Will make specific and open attacks condemning individuals and situations. |
| Underlying question for the facilitator | What is the end, bottom line result they want? |
| Facilitator tactics | Assertively focus on results. Concentrate on discussing the realistic actions you are taking without being bullied. |
| Facilitator goals | Be treated with respect. Be valued for what you can do and be allowed to get on with it. |
| **EXPLOSIVE** | |
| Description | Frustration simmers and will explode at minor or unrelated issues. |
| Typical behaviours | Will appear to 'go off on one' on occasions where blame is aimed at everyone and everything. |
| Underlying question for the facilitator | What is their real issue with the situation? |
| Facilitator tactics | Don't try to tackle the issue until they have calmed down. Ask open questions about what makes them angry. |
| Facilitator goals | Identify specific issues and bring issues out into open where they can be dealt with. |

## RUDE

| | |
|---|---|
| Description | Tries for smart one-upmanship. |
| Typical behaviours | Chips away using sarcasm or pointed remarks. Typically will play to group. |
| Underlying question for the facilitator | What is their real issue with the situation? |
| Facilitator tactics | Positively question the relevance of the comments. Ask for the underlying reason for their comments. |
| Facilitator goals | Bring issues out into open where they can be dealt with. |

## KNOW-IT-ALL

| | |
|---|---|
| Description | Knows their stuff but can only see narrow perspective. |
| Typical behaviours | Often technically very capable, they have very clear views on issues and what the right solution is. |
| Underlying question for the facilitator | How do I show value in their expertise as *part* of the solution? |
| Facilitator tactics | Don't become a 'know-it-all-better' but know your stuff. Sincerely value their contribution and offer to share progress and ideas. Assert responsibility to look fairly at new ideas. |
| Facilitator goals | Be treated with respect but also emphasise the respect you have for them. Move to an adult conversation about possibilities. |

## THINKS-THEY-KNOW-IT-ALL

| | |
|---|---|
| Description | Likes attention and will talk with great conviction but little fact about an issue. |
| Typical behaviours | Will tend to try to dominate opinions with sweeping judgements or opinions. Is liberal in interpreting actual data. |
| Underlying question for the facilitator | How do I challenge their assumptions without challenging them? |
| Facilitator tactics | Thank them for their contribution and reaffirm the objectives of the discussion. Without malice, ask for specifics. Offer to share your evidence with them as a way of catching and focusing their enthusiasm. |
| Facilitator goals | Focus on facts and realities without humiliating others. |

▶

## NO DELIVERY

| Description | Will agree to actions but appears unable to follow through. |
| --- | --- |
| Typical behaviours | At first, they may appear helpful, but despite regular approaches appears incapable of completing agreed longer-term tasks. Highly reactive to here and now demands. |
| Underlying question for the facilitator | How can I show they will truly be valued by what they can deliver? |
| Facilitator tactics | Work to encourage and value honesty. Be supportive in showing the consequences of non-delivery. Make it easy for them to plan and feed back progress. |
| Facilitator goals | Get commitments that the person feels strong enough to protect and defend. |

## THE PROCRASTINATOR

| Description | Will seek to delay a decision until it has almost made itself. |
| --- | --- |
| Typical behaviours | Will not give a straight answer or will require more information or research on a variety of issues. Will often ask for more time. |
| Underlying questions for the facilitator | What is the underlying block for the person? What is stopping them moving forward? |
| Facilitator tactics | Create a comfort zone that encourages sharing of concerns. Encourage adoption of a decision-making process (which includes acknowledgement of risk). Having worked jointly to a conclusion, pass final decision back to them. Reassure confidence and ensure follow-through. |
| Facilitator goals | Gently encourage confrontation of risk and uncertainty in a supportive and structured way. |

## SILENT TYPE

| Description | Appears unreactive to people or situations. |
| --- | --- |
| Typical behaviours | Rarely comments, shares ideas or complains. Will tend to be monosyllabic in their responses and will give little away when asked. |

| Underlying question for the facilitator | How can I get this person to want to talk? |
|---|---|
| Facilitator tactics | Relax and don't hassle. Create a comfortable non-pressure environment to talk. Ask open questions, use silence and ask positively again. Make supportive guesses on reasons for not talking. Look for any clue and build on it. |
| Facilitator goals | Create a safe enough environment where individuals will bring issues out into open where they can be dealt with. |

## IT WON'T WORK!

| Description | Always looks at the negative side of the situation. Sees little point in activities and changes. |
|---|---|
| Typical behaviours | Will say 'no' at the first opportunity. Will point out faults and problems without any attempt to offer solutions. |
| Underlying question for the facilitator | How do I move from problem identification to outcome/solution? |
| Facilitator tactics | Empathise – recognise the value of truth behind their generalisation. Use them as a resource to specifically challenge ideas. Don't rush them but allow time to consider options. Build from here. |
| Facilitator goals | Welcome the value of criticism to uncover risk and uncertainty – from which solutions can be sought. |

## HELPLESS

| Description | Will complain about situations and circumstances. Sees themselves as a victim. |
|---|---|
| Typical behaviours | Likes to share problems and difficulties which appear to result from someone or something else. |
| Underlying question for the facilitator | How do I move the perception of helplessness to ownership or responsibility? |
| Facilitator tactics | Listen and write down the main complaints. Get specific information on each issue. Ask what they want in each situation. Agree a plan of action given the facts of the situation. |
| Facilitator goals | Create a safe environment where issues come out into open. Firmly switch focus to a solution-based discussion. |

## 🡕 **brilliant** exercise

Which type of person is your most challenging? They may not be on the above list. Use the questions below to help you prepare for your next workshop.

Consider the following points and jot down your thoughts:

● Why do you think somebody would behave like that?

● What is it about their behaviour that would make you feel uncomfortable?

● What impact might this person/behaviour have on the rest of the group?

● How might this person best be dealt with?

## **brilliant** recap

● Set your ground rules – at any time.

● Remember participant names by repeating them back on introductions.

● Know which stage of Tuckman's cycle your audience is at – and adapt accordingly.

● Create exercises for your Adaptors and Innovators, as well as for the four social styles of Bolton and Bolton.

● Deal with individuals by understanding their reasons for behaving like they do.

**CHAPTER 10**

# How well is it going?

This chapter will look at checkpoints for you to ascertain how well your workshop is going. Many facilitators say, 'I *think* it's going well...' But you need more than a hunch, as at the end of the day it's too late to rectify anything. We will look at the good practice of handing out evaluation forms at the outset and how this feedback can be your biggest friend. Beware the after-lunch/mid-afternoon drop in energy. People will remember your workshop if it ends on a high. Don't fall into the trap of letting the day run down to the end as people 'seem' weary. As a brilliant workshop facilitator, you are there to challenge and to engage with the audience. You are in control and you have to maintain the positive environment effective for learning.

 The purpose of education is to replace an empty mind with an open one.

Malcolm Forbes

'I *think* it's going well.' I have lost count of the amount of times I have heard this response when I ask other facilitators how they think the session is going. Of course, we have strong instincts that can guide us, but we can never be really sure. The word 'think' in that sentence should signal to us that we need more concrete evidence to give us confidence to continue as we are doing.

The big danger of not checking in with your audience until the end is that it is too late to rectify the situation. Just like a manager having an appraisal review with their member of staff and informing them of something that could have been done better a few months ago!

Although evaluation will be covered in the next chapter, it is worth mentioning that you should use a form of regular feedback throughout the workshop to enable you to adapt your approach if required. The danger of just asking the group how it is going is that they will respond with 'fine' or 'great'. While these are positive words, how can you replicate 'fine' or 'great' throughout the workshop?

Let the group know that you will be asking for feedback throughout the workshop. Tell them you will be asking them about their energy levels, their thoughts on the exercises undertaken and any feedback for you as the facilitator.

One friend of mine asks the group at the beginning of the introductions session to complete the sentence, 'I hope the facilitator is…' This enables her to get a feel for the type of style the group wants. Obviously, this may be different from what they actually need, but it normally adds value to her. However, you need to have some experience behind you – and a little courage – to use this approach.

Sometimes asking the whole group how things are going can be ineffective. Try asking a group if everything is 'going OK'. More often than not you will be met with a few nods and grunts. Unless you are having a severe off-day, it is very rare that someone will speak up and say that it is not going well – although it can happen.

## Cultural architects

The ex-England football manager, Sven Goran Eriksson, identified David Beckham as one of his 'cultural architects'. Cultural architects are defined as people who are able to change the mindset of others. As Sven couldn't be with the team the whole time, he chose a few players who would be his eyes and ears and share the same vision of success as he did.

In the same way, it can be useful to build rapport with certain individuals and use them to understand how the workshop is going and what others are saying about it. If you know in advance that certain participants support the reasons for the workshop, you can ask them beforehand if you can use them as one of your cultural architects.

**brilliant tip**

Give out your evaluation form at the beginning of the workshop, not the end. Get participants to complete it as they go through the sessions and leave them on the desks. That way you can walk round and take a look to gauge current, not retrospective, feedback, and act on it.

## Energy levels

Energy levels also need to be measured. At the beginning of the session there may be a lot of energy, whether due to nerves or anxiety. And this may just be the facilitator! The group energy levels need to be considered. The general rule of thumb is that at certain points of the day there are differing levels of energy (see Figure 10.1 overleaf).

Although this chart is not based on deep research supported by statistics, it is based on 17 years of experience in workshops, with supporting evidence by colleagues. It is meant only to be a guide to enable you to think about the possible differing levels of energy and the type of interventions needed to keep the group engaged.

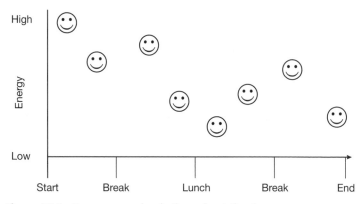

**Figure 10.1** Group energy levels throughout the day

At the start, energy levels are generally high. The facilitator sets the tone through positive energy and the group are ready for the first exercise. The beginning should be a high impact intervention.

As the first break draws near, minds might be on that cup of tea or coffee, or that cigarette or email that they have been waiting for. Expect a slight dip in energy levels or commitment to the exercise because of this – but not always.

Refreshed after the break, get the group ready to go again. Another high impact session will be matched by still quite high energy levels.

As lunch draws near, again minds will be on the break. If lunch is brought into the room, it can be a huge distraction as it is the signal that the food has arrived and everyone is eager to see what is on offer. Be aware that people will see lunchtime as a key milestone in the day. They may have made plans or they just may be starving. Do not keep them longer than necessary. Have lunch earlier than planned if the group are sidetracked.

**brilliant** tip

If possible, make sure the lunch is left outside the room to avoid distractions.

The after-lunch session is where energy is at its lowest. Also known as the 'graveyard shift', participants have full stomachs and are ready for a 30-minute nap. Begin with an interactive session that requires all of them to be involved. This could be as a group or in small groups or pairs. Do not begin the afternoon session with a long talk during which all they have to do is listen. Watch for the eyes closing...

The energy levels will increase gradually after lunch so keep the group active. The afternoon break will be a last time to refuel and really focus on the remainder of the session. Get across your key messages and engage as much as possible.

As the end draws near, individuals may disengage. Watch for body language – sitting back in the chair, yawning, looking at their watch or eyes getting glazed. Energy levels are getting lower as the end of the day approaches. Go out on a high. Never prolong the session if you feel that it will be a long, drawn-out affair. If you have covered everything you need to, finish earlier at a high point to end the workshop on a positive note.

Remember, this is only a guide. All groups are different. You will need to manage the energy levels. Always make sure your own levels are high throughout. You are the role model for energy. Let them feed off your energy.

 **brilliant** recap

● Begin with a high and end on a high. People can usually recall the beginning and the end.

● Keep your levels of energy high throughout.

● Manage group energy levels, especially after lunch and towards the end.

● Identify and choose cultural architects to help you.

# PART 3

# Encore (post-workshop)

W ow! How good was that? You've come off stage and you just want to relax. They were a tough audience but you managed to leave them wanting more. You've run out of time on this occasion as the house lights are already up and the room is emptying. How do you get to return?

## How do I measure my effectiveness?

Chapter 11 looks at formal evaluation of your session. It is a necessary evil for many organisations but it needs to be a crucial part of the wider learning process. We look at when to evaluate and when not to.

## How did my workshop add longer-term value?

Chapter 12 gives you tips to keep learners engaged long after the workshop. Learning is a lifelong process and to add real value for your clients you need to do all you can to make sure that the learning is transferred back into the workplace. This is their real return on investment.

Lastly, we take a light-hearted look at some facilitator tales, where these people had a tough time and wanted to close their eyes and wish it never happened…

**CHAPTER 11**

# Evaluation

I n this chapter we will look at the different types of evaluation
form (also known as 'happy sheets') and their use and appli-
cability, linked to Kirkpatrick's famous evaluation model. We
will identify when to use them and when not to; and examine
why some people are reluctant to complete them and how we
can challenge their mindset.

 Trainers must begin with desired results and then
determine what behaviour is needed to accomplish
them. Then trainers must determine the attitudes,
knowledge, and skills that are necessary to bring about
the desired behaviour(s). The final challenge is to
present the training programme in a way that enables
the participants not only to learn what they need to
know but also to react favorably to the programme.

Donald Kirkpatrick, 1998

It is that time of the day, when the evaluation forms are handed out
and the facilitator either tries to make small talk or just keeps very
quiet, waiting with bated breath to see if it has all been worth it.

As mentioned in the last chapter, evaluation forms can be given
out at the beginning and completed as the workshop develops.
However, on the whole, forms tend to be handed out at the end.
The reality is that most people want to get away at the end of
the day to catch their train or bus. The evaluation form is rushed
through, and a few ticks here and there aren't given much thought.

Now, I don't know about you, but when I have worked hard all day to enable the group to have a positive learning experience, I expect a little more focus on a process that should be given higher priority than many organisations give it. It is mainly the facilitator that looks through the ticks and comments and breathes a sigh of relief when there is nothing really challenging in the feedback.

Nothing great was ever achieved without enthusiasm.

Ralph Waldo Emerson

## To evaluate or not?

That is the question… An evaluation takes time and can cost money. The cost can be attributed to the time spent by an individual to collate and enter all the data onto a spreadsheet, for example. A decision needs to be made whether it is worthwhile to evaluate at all.

The greater the cost of the workshop, the more important it is to evaluate the effectiveness of the decision to spend the amount and the value that the workshop has given to the participants. If the workshop is likely to be repeated, then again, it is worthwhile to evaluate so that improvements can be made if necessary.

He that will not reflect is a ruined man.

Asian proverb

Also, how many people will be attending the workshop? If it is high in proportion to the size of the organisation, then it is important that value for money is assessed.

A colleague showed me a model she uses to help her decide whether to evaluate a particular session or workshop. There are a range of factors that are scored on a scale of 1–10 (see next page). If the total score is more than 25 it will be worthwhile evaluating the session or workshop. If a particular factor scores 7 or more then think about whether that particular factor should be assessed in greater detail.

Low level of investment                           High level of investment

| 1 | 2 | 3 | 4 | 5 | 6 | 7 | 8 | 9 | 10 |
|---|---|---|---|---|---|---|---|---|----|

Small percentage of staff        High percentage of staff involved
involved

| 1 | 2 | 3 | 4 | 5 | 6 | 7 | 8 | 9 | 10 |
|---|---|---|---|---|---|---|---|---|----|

One-off workshop                                 Repeat workshops

| 1 | 2 | 3 | 4 | 5 | 6 | 7 | 8 | 9 | 10 |
|---|---|---|---|---|---|---|---|---|----|

Low link to business objectives  High link to business objectives

| 1 | 2 | 3 | 4 | 5 | 6 | 7 | 8 | 9 | 10 |
|---|---|---|---|---|---|---|---|---|----|

Familiar learning intervention                   Unfamiliar learning
                                                 intervention

| 1 | 2 | 3 | 4 | 5 | 6 | 7 | 8 | 9 | 10 |
|---|---|---|---|---|---|---|---|---|----|

It really all depends on what the organisation wants measured and/or costed. There are different types of cost:

- Direct cost – these are the costs that are incurred only if the workshop is run.
- Indirect cost – these costs are variable and part of supporting the workshop.

| Direct costs | Indirect costs |
|---|---|
| The venue | Flipcharts/pens |
| An external facilitator (including travel expenses) | Projector |
| Refreshments | Administrative support |
| Resources (materials) | Room clearance/cleaning |

Once these costs are included, a workshop's total cost is often doubled.

## Kirkpatrick's model of evaluation

Professor Donald Kirkpatrick (1959) developed the most famous model of evaluation, containing four levels (see Figure 11.1). (See **www.kirkpatrickpartners.com** for further information.)

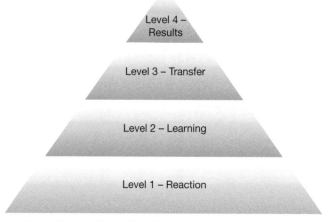

**Figure 11.1** Kirkpatrick's model of evaluation
*Source*: Based on data from Kirkpatrick (1959)

### Reaction

This is normally the reaction of learners immediately after the workshop. They complete the evaluation forms based on content, objectives, the facilitator's effectiveness and resources or materials used.

| Pros | Cons |
| --- | --- |
| Provides immediate data. | Provides data on how people 'felt'. |
| High completion rate if completed on the day. | Data can be rushed. |

## Learning

This stage measures what people have learnt. To measure this effectively there needs to be some pre- and post-workshop questionnaires completed, to calculate any change in attitude, improvement in knowledge or reduction in skills gaps. Some feedback from their manager/colleagues after the event can give further data on whether the individual has added value to the team/department since the workshop.

| Pros | Cons |
| --- | --- |
| Measures if the new skill or knowledge has made a positive difference. | Collating the post-workshop evidence can be time consuming. |
| Can validate the success of the learning intervention. | May not be able to apply the workshop learning environment back into the workplace. |

## Transfer of learning

Normally undertaken three to six months after the workshop. This measures how the learner has transferred the learning into the workplace after a longer time period.

| Pros | Cons |
| --- | --- |
| Can really measure the impact of the workshop with detailed data. | Can be costly. |
| Line manager role crucial – buy-in here can add longer-term value in development. | Can take a large amount of time to complete and validate the data. |

## Results

This measures the impact of the workshop on organisational effectiveness. It could be that the organisation has increased sales, reduced complaints in customer service or managed staff performance more effectively. This stage is very difficult to measure. Correlating an organisation's success with a workshop or workshops is not easy as there may be several other contributing factors.

| Pros | Cons |
|------|------|
| If there is senior management buy-in then learning is high on the agenda. | Getting buy-in from senior management. |
| Can provide an organisation with valuable return on investment data. | Very time consuming. |

## Other models of note

Although there are many models of evaluation, the two that I grew up with, apart from Kirkpatrick, were the following:

1  The CIRO model (Warr, Bird and Rackham, 1970).

2  Five level evaluation (Hamblin, 1974).

### The CIRO model

| Context | Inputs | Reactions | Outcomes |
|---------|--------|-----------|----------|
| Were the needs analysed accurately at the outset? Why was this particular workshop chosen? What were the learning objectives? Does it link to the organisation's culture? | What resources were available to support the workshop? Did we choose the right participants? | Did the workshop achieve what it set out to do? (Feedback from the participants.) | Did the outcomes match the original objectives? Has the workshop been a success? |

*Source:* Based on data from Warr, Bird and Rackham (1970)

## Five level evaluation

| | |
|---|---|
| Level 1 – reactions | What did the participants think of the workshop? |
| Level 2 – learning | Did the participants learn what was intended? |
| Level 3 – job behaviour | Has the participant transferred the learning to the workplace? |
| Level 4 – organisational/departmental | Has the workshop assisted organisational/departmental goals? |
| Level 5 – ultimate value | Has the workshop impacted on profit/customer service/public perception? |

*Source:* Based on data from Hamblin (1974)

# Evaluation examples

There are a number of ways that you can collect information to evaluate your workshop. For this book we will be focusing on evaluating the workshop on the day – level 1, reaction – so that you can design a questionnaire that works for you. Ideally line managers should be involved in the whole process but, in my experience, this rarely happens. We must presume that they don't want to be or that the organisation does not think it is necessary.

The forms can vary depending on the organisation and its culture and the type of workshop you have run. If you do get a say in the design of an evaluation form, always attempt to tailor it specifically to the subject. Generic forms are used by organisations for consistency in collating the data, but sometimes you cannot measure like with like as the subjects are so different.

Some forms allow you to circle a few words about how you felt about the workshop (see example overleaf).

difficult          stimulating          interesting

fun          confusing          waste of time

helpful          boring          superficial

rewarding          constructive          instructive          exciting

This is acceptable as long as it is followed up with questions unpacking what they meant by their selection. Too often, participants circle a word and there is no follow-up question. (Refer to the open and closed questioning skills in Chapter 1.)

There are many forms with an opening question of 'What were your personal objectives before attending the workshop?' This will not work unless the organisation has a clear learning strategy and each individual has sat down with their line manager to have this discussion. This rarely happens. This therefore makes the question redundant but organisations still ask.

Some styles prefer the participant to have free rein to write just how they think the workshop benefited them. The example below is a simple four-quadrant model on an A4 sheet.

| What I have learnt today | How it applies to my role |
| --- | --- |
| What I am going to do differently | Follow-up action |

The design of your questionnaire can be a mix of open and closed questions, usually with some rating scale questions.

## Examples of rating scale questions

How useful was the information received today? (1 is high, 5 is low)

| 1 | 2 | 3 | 4 | 5 |
| --- | --- | --- | --- | --- |

What is the problem with this question and style?

Too often, the rating headings are not defined. For example, if the rating headings are 'excellent', 'good', 'satisfactory' and 'poor', it is going to be open to interpretation as a 'good' for one person may be a 'satisfactory' for another. Defining the levels is important.

It is the same when there is just a number – usually stated as, 'If 1 is high and 5 is poor, complete the following assessment.' It is a very personal scoring system and more clarity should be given, as everyone has different standards.

When asking how useful was the information received, there is no clarity around the words 'useful' and 'information'. Are people measuring the same thing? Probably not.

At level 1, we are really seeing the reaction to the workshop and what people have learnt. We tend to capture feedback on the admin processes, the lunch, the room, the facilitator, the hand-outs, etc. For example, what has just happened? How was the experience for you?

Three styles that I particularly like are as follows. They have been used alongside other evaluation forms but they work for me – and, more importantly, the participants and the organisation.

## Stop/start/continue

This is very simple. Each participant is given an A4 sheet at the beginning of the day with three sections split horizontally. They complete it as the workshop progresses.

They make notes after each session as to what they have learnt by completing the sheet.

1   What are they going to *stop* doing (or do less of) after completing that session?

2 What are they going to *start* doing (or do more of) after completing that session?

3 What are they going to *continue* to do – as the session showed them that they are doing it well?

## Pre- and post-workshop

The second form that is very effective for measuring learning gaps is one that is given out at the start. They complete one section before the workshop commences and then complete another at the end, using their own rating scale.

The following example is taken from a 'Managing Change' workshop:

● Complete column A *before* the workshop commences.

● Complete columns B and C *after* the workshop finishes.

● Add up column totals and compare pre- and post-workshop ability.

0 = lowest and 10 = highest

|  | Column A (pre-workshop) | Column B (post-workshop) | Column C (change between columns A and B), +/− |
|---|---|---|---|
| I am aware of the impact of change on my team | 6 | 8 | +2 |
| I am aware of how I personally react to change | 4 | 7 | +3 |
| I am confident I can deal with the upcoming restructure | 2 | 7 | +5 |

This can show a learning shift for a participant. It can also show how effective the workshop has been is helping others to attain the skills required to implement – in this case – the upcoming changes.

## Commitment plan

The last choice is called a 'commitment plan' – this idea was formulated in 2004 by W. Leslie Rae and full details of the plan can be seen at **www.businessballs.com/trainingevaluationtools. pdf**. At the end of a workshop a commitment plan should be completed based on what has been learnt or has been reminded. When learning is applied on their return to work, the new skills and knowledge develop, reinforcing new abilities, and the organisation benefits from improved performance. Learning without meaningful follow-up and application is largely forgotten and wasted. The plan is in two parts.

Firstly, participants complete three to five key areas of learning from the day. An example is shown in the table below.

|   | Item | How to implement | When I will implement |
|---|------|------------------|------------------------|
| 1 | Be more assertive in managing the team | Make decisions and stick with them | In the next month |
| 2 | Improve my communication skills | Get feedback from other managers | Next week |
| 3 | To listen more effectively | Get feedback when I talk over people | Commencing immediately |

The second part of the plan is to take each one individually and then to build an in-depth approach to achieving your commitments. Obviously this takes time, so get participants to complete at least one action point at the workshop and then complete the others with their line manager back in the workplace.

Questions that need to be answered in the second part are given in full at **www.businessballs.com/trainingevaluationtools.pdf** (page 7).

Whatever the reasons you and/or the organisation may have for evaluation, make sure that you are measuring the right things. You need to know the purpose of the training, its context and the objectives before any evaluation methodology can begin. Ask why the evaluation is taking place. What will it measure? How will it be measured? What are you going to do with the results?

## Convincing participants

All too rarely do we make the participants understand the importance of evaluation and their role in helping the organisation meets its challenges of the future. By having a short session at the outset on evaluation and its importance, you have more chance of getting buy-in. However, it is challenging to get a positive message across when the culture is seen as one that does nothing with the evaluation once collated. It is just a process to them.

Few people see the results of the evaluation outside HR or L&D departments. The participants have completed the forms and yet they will be the last to know – or they will never know!

With advancement in technology accelerating at increasing speed, many evaluations are completed online. Participants have an email waiting for them after the workshop and they have to complete it within a certain timeframe. The data can be collated and put into a report within seconds and the organisation can evaluate the success of the workshop.

**brilliant** tip

Once the evaluation data is collated, as well as sending it to HR or another department in the organisation, send the report to the participants who completed it in the first place. It is courtesy to do so and elevates the priority an organisation gives to evaluation.

 **recap**

● Identify whether it is beneficial to evaluate the workshop.

● If so, know what it is you are measuring and why, and at what level.

● Evaluation begins when the workshop starts.

● Use a mix of open/closed questions and rating scales.

● Always define rating scales with definitions.

● Share the evaluation results with the participants.

**CHAPTER 12**

# What next?

t's been a month now since you delivered that workshop. How are people doing? Has it made a difference? In this chapter you will learn about post-workshop prompts and learning nudges that will ensure your participants haven't forgotten how valuable the learning has been. The real value is in people applying their learning in the workplace and trying out new things for the benefit of the organisation.

 If you think training is expensive, try the cost of ignorance.

Tom Peters

The level 3 and level 4 evaluation stages (see Chapter 11) will have an impact a few months after the workshop. You will need to work with line managers to have the discussions or, again, if the technology is in place, an online system can be effective.

Questions can be asked around:

- How successful were the participants in implementing their action plans?

- To what extent were they supported in this by their line managers?

- To what extent has the action listed above achieved a return on investment (ROI) for the organisation, in terms of either identified objectives satisfaction or, where possible, a monetary assessment?

The most valuable ROI for any organisation is the difference individuals will make by applying their learning and putting it into practice in the workplace. They will benefit, their team will benefit and their manager will benefit, as well as the organisation.

As we have discussed, depending on the organisation's culture, this is a very difficult step to put in place. The workshop in question happened about a month previously and now we need to know the real value of what was learnt.

## Email prompts

A brilliant way of ensuring that the learning is not forgotten is to use email prompts regularly after the workshop. I often offer this to clients as a way of embedding the learning at no extra cost to the client.

After the workshop – usually a week after – a series of weekly emails are sent to each participant automatically, with each one reminding them of the key learning objectives and offering tips and hints to encourage application. The emails are set up to be sent automatically to each participant and it is very easy to administer.

### brilliant example

A client had received a customer service workshop focusing on improving response times to phone calls and emails. Each week for 10 weeks, each participant received an email at the same time on a Monday morning, with each one having a theme carried over from the workshop.

The emails reminded them of top tips – how to respond appropriately to telephone calls and to understand the assessment techniques used by the organisation that would measure success. Managers were aware of these emails and were in constant communication with the participants so that customer service was always on the radar. This approach was well received and it made a big impact, assisting them to succeed in achieving their goals.

The positive use of new skills, behaviours or learning cannot be underestimated. As a facilitator, we should be pushing this approach to our clients and making them see the value to their organisation. To correlate success of the organisation with a particular learning intervention is very difficult at the best of times, so evaluation should be a core part of the L&D strategy.

## Learning nudges

Learning nudges are commonly used by many organisations and are similar to the previously mentioned email prompts. These may be emails, but could also be small articles placed on their intranet site, or bite-size learning activities, or stories about an individual successfully implementing their new skills.

### Articles

These are very useful for individuals who like to read and absorb information. They may be about a particular theory or a more in-depth view of a model or tool used at the workshop.

### Bite-sized learning

These can be individual topics in their own right, or they can be additional exercises that supplement the workshop. The advantages are that some people prefer to learn in this way, it is cost-effective as there is no need to leave your desk, and it can enable individuals to manage their learning more effectively. The downside is that, although there are positive intentions, if the culture is not right for taking the time to learn then the exercises and articles will just sit on the server.

E-learning is now a fundamental consideration in an organisation's L&D strategy. It can be stand-alone or it can support other learning interventions. Work with organisations to find out their learning provisions and align your evaluation to it.

## Just pick up the phone!

If the technology in some organisations does not support evaluation – as it can be costly – then how about just picking up the phone? This is an extremely underused method but it is extremely effective.

### brilliant example

At the outset, we agreed with a client that the evaluation would include a telephone call with all the participants a month after the workshop. A set of questions were agreed in advance and the participants were informed that this would happen at the end of the workshop and to expect the call.

Each person then had a discussion with the facilitator at an agreed time. The benefit of a conversation is that questions could be followed up and more information gained – unlike that from a set questionnaire.

Example questions asked of the participant:

● Has there been an improvement in your performance at work?

● What percentage of your improvement can be attributed to the workshop? What is the basis for your percentage? (Get examples.)

● What other factors may have contributed to your improved performance?

To add value and to assess the accuracy of the participant's feedback, build in telephone conversations with their line manager. You will have a large array of data that you can present back to your client to prove their investment in the workshop was the right choice.

## A letter to me

This approach can also make a positive impact after the workshop. At the end of the workshop, get everyone to complete their action plan in the normal way. Rather than let them leave with it, give out envelopes so that they can write their name/department and address (if applicable) and then get them to place the action plan in the envelope and collect them in. Whether it is a day/week/month after the workshop – whatever will have the most impact – post back their letters to them. This will act as a reminder of what they have committed to do and will bring back the memories of the workshop.

## brilliant recap

- Real ROI for an organisation happens when their staff put their learning into practice in the workplace.

- Keep that workshop in their minds by using email prompts at regular intervals.

- Use learning nudges to supplement learning.

- To know how people are doing, pick up the phone. Speak to participants and their managers.

- Capture success and spread it!

# Conclusion: Did that really happen? I don't believe you!

Here follow real-life stories from trainers and facilitators about times when they wished the ground had swallowed them up. As the saying goes, 'We learn far more from our mistakes than from our successes.' Try telling it to these people! By following the advice in Chapters 1–12, this won't be you.

 While one person hesitates because he feels inferior, the other is busy making mistakes and becoming superior.

Henry C. Link

What could go wrong? The planning and design have been completed and we are feeling confident we have the tools and techniques to help the participants have a successful day. Well, things can go wrong, and it is OK to make mistakes, but learn from them. Failure is not falling down but staying down.

Anyway, here are a few tales from colleagues about themselves and others who have experienced that moment when they wished the ground would swallow them up. They would like to remain anonymous... for obvious reasons.

'I was speaking at a conference and had a microphone as it was quite a large audience of about 250. Although I was a confident speaker this was the first time I had needed a microphone. Anyway, it was all going well and the break came and I thought I'd just leave the room and focus on the next session. I checked my phone and saw a missed message from my mother, so I called her. We chatted about the usual stuff, as well as my upcoming birthday and what I wanted. When I mentioned my husband was buying me lingerie I heard a large noise come from the conference room. I thought nothing of it until a participant came rushing out and told me my microphone was still on and they had heard every word!'

'I was attending a colleague's presentation to offer support. He was quite relaxed and had delivered many times before. I remember it was a hot afternoon and there were about 10–12 people in attendance. To be honest, he is not the most dynamic speaker, and I think he knew that so he put on a video for the group to watch and make notes and then we would complete an exercise. The lights were dimmed and everyone was writing down their thoughts. When the video was finished I thought I'd be helpful and get up and turn it off. I did so, as well as turning the lights back up. However, my colleague was fast asleep! To say he was mortified was an understatement. My rule – never fall asleep in your own presentation!'

'This could surely have only happened once. I am a very careful planner and always have contingency plans just in case I can't rely on the equipment. On this occasion I was showing a presentation on the organisation's room projector. As you sometimes fear, the projector was not working on the day. Never mind, I had hard copies of the slides on the table next to me. Not so apparently.

When I went to the toilet earlier a cleaner had come in and, thinking they were left over from the previous evening, had thrown them out. Still, I was in control as I had brought laminates that I could show on the overhead projector – if it wasn't being used by the boss! I remember just talking through the session. It wasn't very memorable. Not one of my better ones.'

'I remember turning up at a hotel to run a session on time management to approximately 10 managers. The obvious thought is that the management of time is better controlled when we always know what the time is. Therefore, as an exercise, I asked all the participants to remove their watches. They then had to complete an exercise in small groups in breakout rooms while they left their watches in the main room. I wanted them to work without time constraints and just focus on the task in hand. To cut a long story short, the watches ended up being stolen as I mingled with the groups. That is what you call a memorable session.'

'I was a witness to this one. I was at a government department and was in the audience so remember it well. We were listening to our new three-year strategy and waiting for our new chief executive to launch it. We had never met him as he had only started the day before – as happens in government departments! Anyway, in strode this man and immediately started speaking to us about his new book. He was proud that we were supporting his ideas and was so happy to see that we would actively be pushing it to the marketplace. It's a shame he was stopped and told he was in the wrong room – I found him to be entertaining. Then we had to listen to our strategy…'

'I turned up to speak at a gym group's staff training day. My friend who worked there and who got me the gig said that I needed to wear their "uniform" of T-shirt, shorts and running shoes and supplied me with the outfit. I arrived and was quickly led into a back room where I met my 'friend' who was wearing the uniform. I changed into mine and was ready. He led me to the back of the stage and told me to go on. Needless to say I was stitched up – as I walked in, every man was in a suit and every woman in a posh dress. I could hear him laughing and I felt helpless, but apparently they do it to all the speakers. For that split second I wanted to crawl under a rock and stay there.'

'Does it count when I designed a supporting workbook for a company, had 100 printed off and had their biggest competitor's logo on the front?'

 'Learning is a treasure that will follow its owner everywhere.'

Chinese proverb

So, now you have a range of tools and tips to help you in the world of workshops. With these at your disposal, I hope that these awkward situations won't happen to you. Of course, I can't predict the unpredictable!

# References

Belbin, R.M. (2010) *Management Teams: Why They Succeed or Fail*, 3rd edn, Oxford: Elsevier. Also at **www.belbin.com** [accessed 25/07/2011]

Bolton, R. and Bolton, D.G. (1984) *Social Style/Management Style*, New York: Amacom

Cameron, E. (2001) *Facilitation Made Easy*, 2nd edn, London: Kogan Page

Facet5 psychometric tool (1994) **www.facet5.com/US/norman. html** [accessed 25/07/2011]

Hamblin, A.C. (1974) *The Evaluation and Control of Training*, Maidenhead: McGraw-Hill

Honey, P. and Mumford, A. (1982) *The Manual of Learning Styles*, Maidenhead: Peter Honey Publications

Kerzner, H. (2009) *Project Management: A Systems Approach to Planning, Scheduling, and Controlling*, 10th edn, New Jersey: John Wiley & Sons

Kirkpatrick, D.L. (1998) *Evaluating Training Programs*, 2nd edn, San Francisco: Jossey Bass

Kirton, M.J. (1989) *Adaptors and Innovators*, New York: Routledge. Also at **www.kaicentre.com** [accessed 25/07/2011]

Kolb, D.A. (1984) *Experiential Learning: Experience as the Source of Learning and Development*, New Jersey: Prentice-Hall

Marston, W.M. (1999) *The Emotions of Normal People*, New York: Routledge

Mehrabian, A. (1971) *Silent Messages*. Belmont, CA: Wadsworth

Music Works For You (2010) 'Business works for you', **www.musicworksforyou.com/the-workplace/workplace-performance.html** [accessed 25/07/2011]

Rae, W.L. (2004) 'Evaluation of training and learning', **www.businessballs.com/trainingevaluationtools.pdf** [accessed 25/07/2011]

SDI (1998) 'Manage conflicts and improve personal relationships', **www.uk.personalstrengths.com** [accessed 25/07/2011]

Tuckman, B. (1965) 'Developmental sequence in small groups', *Psychological Bulletin* **63** (6): pp.384–99

Warr, P., Bird, M. and Rackham, N. (1970) *Evaluation of Management Training*, London: Gower Press

# Index